Murali Banerjee
Jagannath Mazumdar

Nonlinear Vibration Analysis of Shallow Shells - A New Approach

Murali Banerjee
Jagannath Mazumdar

Nonlinear Vibration Analysis of Shallow Shells - A New Approach

LAP LAMBERT Academic Publishing

Impressum / Imprint

Bibliografische Information der Deutschen Nationalbibliothek: Die Deutsche Nationalbibliothek verzeichnet diese Publikation in der Deutschen Nationalbibliografie; detaillierte bibliografische Daten sind im Internet über http://dnb.d-nb.de abrufbar.

Alle in diesem Buch genannten Marken und Produktnamen unterliegen warenzeichen-, marken- oder patentrechtlichem Schutz bzw. sind Warenzeichen oder eingetragene Warenzeichen der jeweiligen Inhaber. Die Wiedergabe von Marken, Produktnamen, Gebrauchsnamen, Handelsnamen, Warenbezeichnungen u.s.w. in diesem Werk berechtigt auch ohne besondere Kennzeichnung nicht zu der Annahme, dass solche Namen im Sinne der Warenzeichen- und Markenschutzgesetzgebung als frei zu betrachten wären und daher von jedermann benutzt werden dürften.

Bibliographic information published by the Deutsche Nationalbibliothek: The Deutsche Nationalbibliothek lists this publication in the Deutsche Nationalbibliografie; detailed bibliographic data are available in the Internet at http://dnb.d-nb.de.

Any brand names and product names mentioned in this book are subject to trademark, brand or patent protection and are trademarks or registered trademarks of their respective holders. The use of brand names, product names, common names, trade names, product descriptions etc. even without a particular marking in this works is in no way to be construed to mean that such names may be regarded as unrestricted in respect of trademark and brand protection legislation and could thus be used by anyone.

Coverbild / Cover image: www.ingimage.com

Verlag / Publisher:
LAP LAMBERT Academic Publishing
ist ein Imprint der / is a trademark of
OmniScriptum GmbH & Co. KG
Heinrich-Böcking-Str. 6-8, 66121 Saarbrücken, Deutschland / Germany
Email: info@lap-publishing.com

Herstellung: siehe letzte Seite /
Printed at: see last page
ISBN: 978-3-659-11093-1

CONTENTS

List of Figures

List of Tables

Chapter 1
INTRODUCTION

A new approach for the analysis of nonlinear vibration of shallow shells of arbitrary shape is presented here. This new approach is based upon the concept of constant deflection contours on the surface of the shallow shell. The constant deflection contour method has previously been applied to the study of linear vibration analysis of shallow shells of arbitrary shape. In this study, the analysis has been extended using the same concept to study the large amplitude vibration of shallow shells. A number of illustrative examples are included to demonstrate the accuracy of the proposed method.

Leissa in his monograph[1] provides a detailed study of linear, vibration analysis of plates. The application of plates and shallow shells in modern technology makes the classical linear theory often inadequate for rational design. Since physical systems are of dissipative in nature, physical linearity and time-independence of mechanical properties of plates and shells do not offer an adequate description of their behaviour. In many modern applications plates and shells experience substantial deflections when they are induced to vibrate, in some cases the size of the deflections being in the order of the thickness of the structure themselves.

In the linear theory of motion of elastic plates and shallow shells, the strain of the middle surface may be neglected when the deflections are assumed to be small compared to the thickness of the surface. However, in most practical cases, this basic assumption is no longer valid. Indeed, the deflections may have the magnitude of the thickness of the surface. Hence the derivation of governing differential equations exhibiting large deflections needs special attention in such non linear analyses.

The importance of the inclusion of nonlinear effects in problems relating to the strength and stability of flight structures has been made initially by von Kármán[2]. Indeed, the von Kármán theory is widely used to account for the influence of large deflection in plates and shells. In fact, more than half a century ago it was Herrmann [3] who first proposed the nonlinear theory of motion of plates, corresponding to the dynamic analogue of the von Kármán theory.

It is well known that the nonlinear dynamic behaviour of thin shallow shell structures is of much technical importance to designers due to its wide range of applications in many fields of engineering. Containers, tanks, domes etc. are common examples of practical importance of such structures. However, the papers on nonlinear vibrations of shallow shell structures to date are limited in number.

The problems of nonlinear vibration of shallow shells have attracted the attention of relatively few investigators in the past[4-7] Due to the very complicated nature of the basic equations governing the motion of a structure exhibiting large deflection, it has always been a difficult task for investigators to obtain even an approximate solution. Mazumdar in 1970 proposed a new approach, which appeared to be quite suitable for bending analysis of elastic plates of arbitrary shapes, based on the concept of iso-deflection contour lines on the bent surface of the plate[8] This simple but efficient method is best known as Constant Deflection Contour Method or CDC-Method. Subsequently, the same method has been extended to the vibration analysis of plates and shallow shells[9-11]

However, the CDC method has so far been restricted to linear analysis until an attempt has been made to extend it to nonlinear analysis of plates [12-13]. In the present study a similar approach as in [13] is undertaken for extension of the study to shallow shells. This study is therefore regarded as a sequel to earlier studies and deals with the nonlinear vibration of shallow shells using the CDC Method. Some specific examples on nonlinear vibration analysis of shallow shells have been included to show the efficacy of the method, and that the results are in excellent agreement with known results in the literature.

Chapter 2
SOME EXPLANATION ON THE CONSTANT DEFLECTION
CONTOUR (CDC) METHOD

2.1 A Basic Concept

The usual methods for obtaining an accurate or approximate solution of problems of bending of plates are based on using the rough idea of the shape of the deflected surface of the plate being physically compatible with the type of fastening at the boundary, the nature of the surface loads and the geometric shape of the plate. In these methods there arise principal difficulties of the satisfaction of the boundary conditions, which practically remained insurmountable for plates of arbitrary shapes. Hence there arises the need for the formulation of an approximate theory, which might appear quite suitable for studying problems on plates, or shells of arbitrary shape.

With this in mind, Mazumdar developed the CDC Method. When an elastic plate with clamped or simply supported boundary is bent under the action of an external pressure, the corresponding deflection surface of the plate may be described by a family of curves, called 'Lines of Equal Deflection', i.e., lines which are obtained by intersecting the bent plate by planes parallel to the original plane of the plate. In principle, it is always possible to determine the equation of such lines of equal deflection.

The basic concept of the constant deflection contour method may be best explained by considering the transverse vibrations of a plate, referred to a system of orthogonal coordinates $Oxyz$ for which Oz is the transverse direction (positively downward). The horizontal plane Oxy coincides with the middle plane of the plate. Consider such a plate, vibrating freely due to either normal static or dynamic loads. When the plate vibrates in a normal mode then at any instant of time t_θ, the intersections between the deflected surface and the parallels $z = constant$ yield contours which after projection onto z = 0 plane are a set of level curves, $U(x,y) = constant$, called the "Lines of Equal Deflections"[10] which are in fact iso-amplitude contours. The boundary of the plate, irrespective of any combination of support, either clamping or simply supported, is also a simple curve C_U belonging to the family of lines of equal deflections, and without any loss of generality one may consider that $U = 0$ on the boundary of the plate.

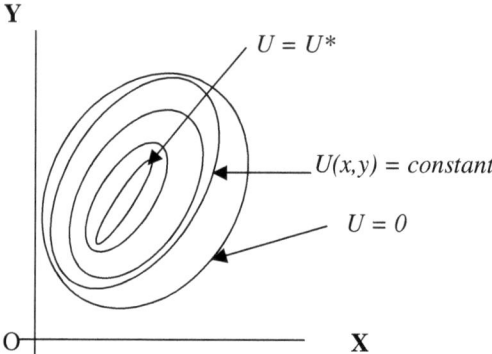

Figure 1 Iso-deflection Curves

As defined in Ref.[10] the family of nonintersecting curves may be denoted by C_U, for $0 \leq U \leq U^*$, so that $C_0 (U = 0)$ is the boundary and C^*_U coincides with the point(s) at which the maximum $U = U^*$ is obtained (Fig. 1).

It is clear that the mathematical form of $U(x,y)$ will differ from one instant of time to the next. However, it is sufficient to consider at a particular instant of time such forces that produce a particular form of $U(x,y)$, since initially solutions for w which are separable in space and time variables are being sought, which will become evident later on.

2.2 Lines of Equal Deflection – U variables

In some cases the equation of lines of equal deflection can be known by symmetry consideration or even by intuition. However, the success of the CDC Method relies on the determination of the 'equation of lines of deflection', as closely as the correct form of the equation.. In cases where the exact equation of the lines of equal deflectioncannot be predetermined, we may find quite good approximations with the aid of an equation. If the boundary of the plate or the shell is given by $A(x,y) = 0$ and the function A is different from zero within the region of the plate, the way to use this equation is to assume a reasonable expression for U of the form[8]

$$U(x, y) = A(x, y) \sum_{\substack{m=0 \\ n=0}}^{m,n} \chi_{mn} \, x^m y^n \tag{2.1}$$

In the linear or nonlinear analysis of plates or shells exhibiting small or large deflections, the use of such lines of Equal Deflection could be assumed. However, for practical interest some known equation of U(x,y) under uniform loading may be cited here.

Table 1 Some known equations for contour lines

Shape of structures	$U(x,y)$
Elliptic plates	$U(x,y) = 1-(x/a)^2 - (y/b^2)$
Annular plate, a and b being the outer and inner boundary respectively	$U(x,y) = a^2 - x^2 - y^2$ $U = 0$, on the outer boundary $U = a^2 - b^2$, on the inner boundary
Equilateral Triangular plates Simply-Supported	$U(x,Y)=$ $[x^3 - 3xy^2 - ax^2 - ay2 + (4/27)a^3][(4/9) a^2 -x^2 - y^2]$
right-angled Isosceles triangular plate Clamped	$U(x,y)=\dfrac{a}{2}(x+y)- xy -\dfrac{1}{2}\left(x^2 + y^2\right)$ $4\dfrac{a^2}{\pi^2}\displaystyle\sum_{n=0}^{\infty}\dfrac{(-1)^n\{\sinh(K_n y)\cos(K_n y)+\sinh(K_x y)\cos(K_y y)\}}{(2n+1^3)\sinh(K_n a/2)}$
Shallow Spherical Shell upon a square base: a = dimension of the base	$U(x,y)=\left(x^2 -a^2/4\right)\left(y^2 -a^2/4\right)$

Chapter 3
DERIVATION OF GOVERNING EQUATIONS

Though our main aim in this study is on the nonlinear vibrating analysis of shallow shells, it would be proper to know, how the CDC method could be introduced to the shell analysis, and what is the advantage in using this method. Initially the method developed by Mazumdar[8] was for the application to various types of problems related to bending and buckling analysis of plates. Later on this method also proved to be efficient enough for problems related to transverse vibrations of shallow elastic and viscoelastic shells[10,14].

We will first discuss briefly one of such problems which will later help our understanding of its application in the forthcoming Sections.

3.1 An Illustration

Consider an elastic, isotropic shallow shell of uniform thickness h subject to a continuously distributed normal load q. Let the equation of the middle surface of the shell referred to a system of orthogonal coordinates xyz, given by Ref. [10]

$$z = \frac{x^2}{2R_x} + \frac{xy}{R_{xy}} + \frac{y^2}{2R_y} \ ,$$

where $r = \sqrt{x^2 + y^2}$ is small compared to the least of the radii of curvature, R_x, R_y and R_{xy} (supposed to be constants).

Consider at any instant τ an element of the shell bounded by any contour line of equal deflection. Since we are primarily interested in the free transverse vibrations of a shallow shell we may neglect the effects of the longitudinal- and latitudinal-inertia terms. Further, if $w(x, y, \tau)$ and $\phi(x, y, \tau)$ denote the transverse displacement and stress function, respectively, then it is possible that

$$w(x,y,\tau) = W(x,y) \cos(\omega\tau + \varepsilon) \tag{3.1}$$

and

$$\phi(x,y,\tau) = \Phi(x,y) \cos(\omega\tau + \varepsilon) \tag{3.2}$$

where $\cos(\omega\tau + \varepsilon)$ is the normal coordinate, ω is the circular frequency, and W and Φ are normal functions determining the form of the defelected surface of the vibrating shell and the stress function, respectively. Since it is the free vibration of shallow shells , we are interested, the application of D'Alembert's principle and summing of the forces in the direction normal to the surface, yields the following dynamical equation (with usual notations):

$$\oint_{C_U} \left(Q_n - \frac{\partial}{\partial s} M_{nt} \right) ds + \int\int_{\Omega_U} [\rho \, h \frac{\partial^2 w}{\partial \tau^2} + \frac{N_x}{R_x} + \frac{N_y}{R_y} + \frac{2N_{xy}}{R_{xy}}] \, d\Omega = 0 \qquad (3.4)$$

Herein the expression

$$[\frac{N_x}{R_x} + \frac{N_y}{R_y} + \frac{2N_{xy}}{R_{xy}}] \, d\Omega$$

represents the net downward contribution of the membrane forces N_x , N_y and N_{xy} acting upon a small element of area $d\Omega$, and the term $\rho \, h \frac{\partial^2 w}{\partial \tau^2}$ represents the inertia force due to the vertical acceleration of the element $d\Omega$, ρ being the mass per unit area of the shell, The double integration being taken over the region, bounded by the contour line $U = Constant$, and the contour integration taken around the closed curve C_U

Using the well-known expression for the moments, shearing forces and the flexural rigidity D and making use of the transformation (as shown in Section 5)

$$\frac{\partial w}{\partial x} = w_x = \frac{dw}{dU} U_x \quad , \quad \frac{\partial w}{\partial y} = w_y = \frac{dw}{dU} U_y \; ,$$

$$w_{xx} = \frac{d^2 w}{dU^2} U_x^2 + \frac{dw}{dU} U_{xx} \; w_{xy} = \frac{d^2 w}{dU^2} U_x U_y + \frac{dw}{dU} U_{xy},$$

Eqn.(3.4) reduces to

$$\frac{d^3 W}{dU^3} \oint_{C_U} R \, ds + \frac{d^2 W}{dU^2} \oint_{C_U} F \, ds + \frac{dW}{dU} \oint_{C_U} G \, ds - \rho \, h \omega^2 \int\int_{\Omega_U} W \, d\Omega$$

$$+ \int\int_{\Omega_U} \left[\frac{1}{R_x} \frac{\partial^2 \Phi}{\partial y^2} + \frac{1}{R_y} \frac{\partial^2 \Phi}{\partial x^2} - \frac{2}{R_{xy}} \frac{\partial^2 \Phi}{\partial x \partial y} \right] d\Omega = 0,$$

$$(3.5)$$

where the factor $cos(\omega \tau + \varepsilon)$ has been cancelled. Here R, F, and G are expressions involving U and its partial derivatives, flexural rigidity D and

Poisson's ratio v. Defining the membrane forces in terms of the stress function and using the condition for the continuity of deformation reduces to

$$\nabla^4\Phi = Eh\left(\frac{1}{R_x}\frac{\partial^2 W}{\partial y^2} + \frac{1}{R_y}\frac{\partial^2 W}{\partial x^2} - \frac{2}{R_{xy}}\frac{\partial^2 W}{\partial x \partial y}\right) \tag{3.6}$$

Thus the problem reduces to solving Eqns. (3.5) and (3.6) for Φ and W given the exact expression for the Lines of Constant Deflection.

To examine the effectiveness of this method the authors in Ref.[10] considered a technically important problem of axisymmetrical vibration of a shallow dome of nonzero Gaussian curvature upon an elliptic base (Fig. 2).

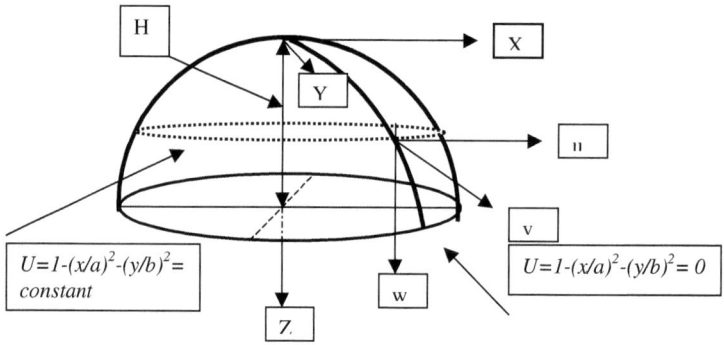

Figure 2 Geometry of the Shell

Although the illustration deals effectively with the determination of the fundamental mode only, the authors mentioned that modes of higher order may also be obtained and the method may be developed further for more complicated problems.

Chapter 4
DERIVATION OF BASIC DYNAMIC EQUATIONS FOR PLATES AND SHALLOW SHELLS

4.1   Governing Equations for Elastic Shallow Shells

As in Section 3.1, we consider an elastic, isotropic shallow shell of uniform thickness h subject to a continuously distributed normal load q. Let the equation of the middle surface of the shell referred to a system of orthogonal coordinates xyz, be given by Ref. [10]

$$z = \frac{x^2}{2R_x} + \frac{xy}{R_{xy}} + \frac{y^2}{2R_y}$$

(4.1)

where $r = \sqrt{x^2 + y^2}$ is small compared to the least of the radii of curvature, R_x, R_y and R_{xy} (supposed to be constants).

If the shell is assumed to be comparatively thin and the displacements (u, v, w) are predominantly flexural, the strain components can be written as

$$\varepsilon_x = \frac{\partial u}{\partial x} + \frac{w}{R_x} + \frac{1}{2}(\frac{\partial w}{\partial x})^2 - z\frac{\partial^2 w}{\partial x^2} = \frac{\sigma_x - v\sigma_y}{E},$$

$$\varepsilon_y = \frac{\partial v}{\partial y} + \frac{w}{R_y} + \frac{1}{2}(\frac{\partial w}{\partial y})^2 - z\frac{\partial^2 w}{\partial y^2} = \frac{\sigma_y - v\sigma_x}{E}$$

$$\varepsilon_{xy} = \frac{\partial v}{\partial x} + \frac{\partial u}{\partial y} + \frac{\partial w}{\partial x}\frac{\partial w}{\partial y} + \frac{2w}{R_{xy}} - 2z\frac{\partial^2 w}{\partial x\partial y} = \frac{2(1+v)}{E}(\sigma_{xy})$$

(4.2)

With usual notations, the total strain energy is given by

$$U = \frac{1}{2}\iiint(\sigma_x\varepsilon_x + \sigma_y\varepsilon_y + \sigma_{xy}\varepsilon_{xy})\,dzdxdy \ ,$$

(4.3)

whereas the kinetic energy is

$$T_e = (\rho h/2)\iint(\dot{u}^2 + \dot{v}^2 + \dot{w}^2)\,dxdy \ ,$$

(4.4)

and the work done is

$$W_k = \iint pw\,dxdy$$

(4.5)

Formulating the Lagrangian with the help of the above expressions and applying Hamilton's principle, a straightforward application of the variational calculus yield the following equations of motion[26]

$$D \nabla^4 w = h\, S(w,F) - h\left(\frac{F_{,yy}}{R_x} + \frac{F_{,xx}}{R_y} - 2\frac{F_{,xy}}{R_{xy}}\right) + q - \rho\, h\, w_{,tt}$$

(4.6)

and

$$D \nabla^4 F = -\frac{E}{2} hS\,(w,w) + E\left(\frac{w_{,yy}}{R_x} + \frac{w_{,xx}}{R_y} - 2\frac{w_{,xy}}{R_{xy}}\right)$$

(4.7)

where the operator $S(w,F)$ stands for

$$S(w,F) \equiv \frac{\partial^2 w}{\partial x^2}\frac{\partial^2 F}{\partial y^2} - 2\frac{\partial^2 w}{\partial x \partial y}\frac{\partial^2 F}{\partial x \partial y} + \frac{\partial^2 w}{\partial y^2}\frac{\partial^2 F}{\partial x^2}$$

Here F denotes the Airy-Stress function, defined by

$$\int_{-h/2}^{h/2}\sigma_{xx}dz = N_x = h\frac{\partial^2 F}{\partial y^2}, \quad \int_{-h/2}^{h/2}\sigma_{yy}dz = N_y = h\frac{\partial^2 F}{\partial x^2}, \quad \int_{-h/2}^{h/2}\sigma_{xy}dz = N_{xy} = -h\frac{\partial^2 F}{\partial x \partial y},$$

(4.8a)

whereas

$$M_x = \int_{-h/2}^{h/2}\sigma_x z\,dz = -D\left[w_{,xx} + \nu w_{,yy}\right] \quad , \quad M_y = \int_{-h/2}^{h/2}\sigma\, z\,dz = -D\left[w_{,yy} + \nu w_{,xx}\right] \quad ,$$

$$M_{xy} = \int_{-h/2}^{h/2}\sigma_{xy} z\,dz = -D(1-\nu)\,w_{,xy}$$

(4.8b)

and the (,) notation signifies partial derivative with respect to the suffix.

4.2 A New Approach

Mazumdar[8] put forward a simple method , the so-called CDC-Method to solve the static and dynamic problems of elastic plates of arbitrary shape. Subsequently, Mazumdar et al [8,9, 14-16] applied this method for solving various problems of elastic shallow shells of arbitrary shape, restricted to linear cases only. Following Mazumdar[8], a new idea has been put forward by

Banerjee[13] to study the dynamic response of plates of arbitrary shape based on the CDC method.

While Mazumdar utilized the concept of deflection contour method to deduce the basic dynamical equations using elementary theory of plates and shells[9] , the authors in Ref.[13] found it easy to arrive at the final equations by straightforward utilization of von Kármán field equations and then utilizing the required transformations to *U-variables*. In most practical cases, it is found that von Kármán field equations in conjunction with the CDC-Method make it easy to apply for nonlinear analyses of plates and shells.

Chapter 5
THEORY AND TRANSFORMATIONS OF EQUATIONS

5.1 Transformation of Basic Equations to U variable

Let $U = U(x,y) = constant$ be a member of the family of iso-deflection or iso-amplitude contour lines. Using the following transformations

$$\frac{\partial w}{\partial x} = w_x = \frac{dw}{dU}U_x \quad , \quad \frac{\partial w}{\partial y} = w_y = \frac{dw}{dU}U_y \ ,$$

$$w_{xx} = \frac{d^2 w}{dU^2}U_x^2 + \frac{dw}{dU}U_{xx} \ w_{xy} = \frac{d^2 w}{dU^2}U_x U_y + \frac{dw}{dU}U_{xy},$$

, etc.

$$(5.1)$$

Eqns.(4.6) and (4.7) can be written as

$$D\sum_{i=1}^{4}\lambda_i \frac{d^{5-i}w}{dU^{5-i}} = h\left[\lambda_5 \frac{d}{dU}\left(\frac{dw}{dU}\frac{dF}{dU}\right) + \lambda_6 \frac{dw}{dU}\frac{dF}{dU}\right] - h\left[\lambda_7 \frac{d^2 F}{dU^2} + \lambda_8 \frac{dF}{dU}\right] + q - \rho h\, w_{tt}$$

$$(5.2)$$

$$D\sum_{i=1}^{4}\lambda_i \frac{d^{5-i}W}{U^{5-i}}hF(t) = h^3\left[\lambda_5 \frac{d^2 W}{dU^2}\frac{d\Phi}{dU} + \lambda_6 \frac{d^2\Phi}{dU^2}\frac{dW}{dU} + \lambda_7 \frac{d\,W}{dU}\frac{d\Phi}{dU}\right]F^3(t)$$

$$+ q - \rho h^2 W\ddot{F}, \qquad \ddot{F} = \frac{d^2 F}{dt^2}$$

$$(5.3)$$

where

$$\lambda_1 = \left(U_x^2 + U_y^2\right)^2 \qquad \lambda_2 = U_x^2\left(6U_{xx} + 2U_{yy}\right) + U_y^2\left(6U_{yy} + 2U_{xx}\right) + 8U_x U_y U_{xy}$$

$$\lambda_3 = 3\left(U_{xx}^2 + U_{yy}^2\right) + 4\left(U_x U_{xxx} + U_y U_{yyy}\right) + 4\left(U_x U_{xyy} + U_y U_{xxy}\right) + 4U_{xy}^2 + 2U_{xx}U_{yy}$$

$$\lambda_4 = \left(U_{xxxx} + 2U_{xxyy} + U_{yyyy}\right) \ ,$$

$$\lambda_5 = \left(U_x^2 U_{yy} + U_y^2 U_{xx} - 2U_x U_y U_{xy}\right) = \lambda_9$$

$$\lambda_6 = 2\left(U_{xx}U_{yy} - U_{xy}^2\right) = \lambda_{10} \qquad ,$$

$$\lambda_7 = \left(\frac{U_y^2}{R_x} + \frac{U_x^2}{R_y} - 2\frac{U_x U_y}{R_{xy}} \right) = \lambda_{11}$$

$$\lambda_8 = \left(\frac{U_{yy}}{R_x} + \frac{U_{xx}}{R_y} - 2\frac{U_{xy}}{R_{xy}} \right) = \lambda_{12}$$

(5.4)

Since Eqns. (5.2) and (5.3) are valid for all points on the surface of the shell, we can have

$$\iint_\Omega \left\{ D \sum_{i=1}^{4} \lambda_i \frac{d^{5-i} w}{dU^{5-i}} - h \left[\lambda_5 \frac{d}{dU} \left(\frac{dw}{dU} \frac{dF}{dU} \right) + \lambda_6 \frac{dw}{dU} \frac{dF}{dU} \right] \right.$$
$$\left. + h \left[\lambda_7 \frac{d^2 F}{dU^2} + \lambda_8 \frac{dF}{dU} \right] - q + \rho\, h\, w_{,tt} \right\} d\Omega = 0$$

(5.5)

and

$$\iint_\Omega \left\{ \sum_{i=1}^{4} \lambda_i \frac{d^{5-i} F}{dU^{5-i}} + \frac{E}{2} \left(\left[\lambda_9 \frac{d}{dU} \left(\frac{dw}{dU} \right)^2 + \lambda_{10} \left[\frac{dw}{dU} \right]^2 \right) + \right.$$
$$\left. - E \left[\lambda_{11} \frac{d^2 w}{dU^2} + \lambda_{12} \frac{dw}{dU} \right] \right\} d\Omega = 0$$

(5.6)

where the integration is over the region bounded by any contour C_U. While performing the above integrals it would be more convenient to utilize the formula in the modified form

$$\iint_\Omega \Psi_1 \left(U, U_x, U_{xx}, \dots, \frac{dw}{dU}, \frac{d^2 w}{dU^2}, \dots \frac{d^n w}{dU^n} \right) d\Omega = -\int_{u*}^{u} \Psi_2(U) \left\{ \oint_{C_U} \Psi_3(x,y) \frac{ds}{\sqrt{\lambda_1}} \right\} dU$$

(5.7)

which is a generalization of the formula adopted in Ref.[10]. Often it has been encountered that in the contour integral in Eqn. (5.7), the integrand turns out to be dependent on U, and hence care should be taken to evaluate first the contour integral. Sometimes, it is useful to use the following relations for evaluation of the contour integral

$$\sqrt{\lambda_1} = U_x^2 + U_y^2 = \frac{4}{p^2}, \quad 2\frac{ds}{p} = \frac{dx}{U_y} = \frac{dy}{U_x}$$

(5.8)

Evaluation of the above integrals will yield two ordinary differential equations. Thus, in accordance with the present method, the basic fourth order partial differential equations reduce to two ordinary differential equations, which make it easier for further study.

Chapter 6
METHOD OF SOLUTION

6.1 A Brief Survey on the Existing Methods

It should be noted here that the above analysis is valid for any shallow shell structure. It has been already stated that the pair of equations (5.5, 5.6) will yield two ordinary differential equations. Yet, the exact solutions of these equations will pose difficult if not impossible task.

Once the basic governing equations are established we will then look for their solutions. The linear approach may sometime look easier than that of nonlinear ones, but sometimes linear problems also involve other geometric nonlinearities or other factors that make the investigation a little harder even for an approximate solution.

Of the existing approximate methods, in general, Rayleigh-Ritz method, Kantorovich method, Finite Element or Finite Difference Method, Galerkin methods, etc are commonly used in the literature. Comparative studies on these methods will allow us to choose the best for our purpose. Elsbrand and Leissa[17] presented the first accurate analysis using twenty-eight terms in the deflection function. The authors raised a very interesting point regarding the trial functions, used in the expression for the deflection function as the product of the beam functions, in order to satisfy all the boundary conditions of the problem. With reference to the work of Kantorovich and Krylov[18] the authors assert that it can be shown that generalised force or natural boundary conditions (such as the free edge conditions) need not be satisfied when using Ritz method. The functional minimisation process of total potential energy guarantees that in the limit the generalised force type boundary condition will be exactly satisfied, together with the governing differential equation of motion. The advantage of satisfying the free edge conditions either exactly or approximately is that the rate of convergence of the minimisation process is increased. The comparison of results shows that the single term Rayleigh method gives reasonably accurate result. Furthermore, from the literature review it appears that the series solution in the form,

Error! Objects cannot be created from editing field codes.

may give higher values of frequency if the fundamental mode alone is not the subject of study.

The above approximate methods have also been extensively used by several researchers for large amplitude vibrations of plates. Vendhan and Das[19] observed that the Rayleigh-Ritz and Galerkin approximations (G-approximations) converge from opposite directions, thus suggesting a possible approach to bind the solutions on either side. In certain cases the sequence of G-approximations may converge faster than R-R method[20]. In the R-R method, it is necessary that the coordinate functions satisfy only the kinematic boundary conditions. However, it is often demanded that the coordinate functions should satisfy both kinematic and natural or force boundary conditions in the application of G-method [21-22].

6.2   Deduction of Time Differential Equation

Therefore, for nonlinear analysis we would apply a Galerkin method and have to seek an approximate solution for which the form of the deflection function *w* must be first assumed compatible with the boundary conditions. Mathematically, this may be explained in the following way. Let *U(x,y) = U* denote a typical member of the family of iso-deflection curves, then for any prescribed boundary conditions the deflection function *w(U,t)* at ant time t can be assumed to take the form

$$w = A \, W(U) \, f(\text{t}) \tag{6.1}$$

where *f(t)* is an unknown function of time to be determined. Using this expression for the deflection function in the resultant equation of (4.6) or in (4.11), we may find the stress function in the form

$$F = \Phi \{U, f(t)\} \tag{6.2}$$

With this expression for the stress function and previously assumed form of *W*, the resultant equation of (5.5) will yield, after using a Galerkin procedure, an *ordinary time differential equation.* Let us suppose that Eqn. (5.5)) in combination with (6.1) and (6.2), will yield the error function in the form

$$\varepsilon = \Lambda^* \left[u, \lambda_1 .. \lambda_{12} \, f(t), \, f^2(t) \, \ddot{f}(t) \right] \tag{6.3}$$

Because of the approximate nature of equation (5.1) , the associated error function may be minimized using Galerkin method. The appropriate orthogonality condition applied to Eqn. (4.5) or (4.10) will yield the following Time Differential Equation with known constants   in the form.

$$\ddot{F}(t) + \alpha_1 F(t) + \alpha_2 F^2(t) + \alpha_3 F^3(t) = q^* , \tag{6.4}$$

the solution of which can be obtained and from which the subsequent analysis can be performed.

Eqn. (6.4) can be studied for the following cases:

(a) Free linear vibration of Plates (when R_x and $R_y \to$ infinity) and shell (as the case may be)

(b) Free nonlinear vibration of plates and shells

(c) Static analysis of plates and shells.

(d) all the above cases (a→c) for elastic-plastic analysis

Chapter 7
SPECIFIC ILLUSTRATIONS

The specific example has been chosen for an illustration with regard to choice of the deflection function as defined in Eqn.(5.1) with a note on judicious choice of the polynomial expression used as the spatial part for the deflection function.

7.1 Illustrations on Plate Problems to Verify the Accuracy of the CDC Method

The present illustrations may be considered to justify the efficacy of the CDC method in solving vibration problems related to plate problems with structures having mixed boundary conditions.

Here we take a different approach without making any deduction of basic equations and using a straightforward application of two field equations extended to a dynamic case. These are well known dynamical equations,

$$D \nabla^4 w = hS(\phi, w) + q - \rho hw_{tt}$$

(7.1)

$$\nabla^4 \phi = -(E/2)S(w, w),$$

(7.2)

in which
$$D = \frac{Eh^3}{12(1-v^2)}, \qquad \frac{\partial^2}{\partial x^2} + \frac{\partial^2}{\partial y^2} \equiv \nabla^2$$

(7.3)

and the operator S is defined by
$$S(I, J) = I_{xx}J_{yy} - 2I_{xy}J_{xy} + I_{yy}J_{xx}$$

(7.4)

It should be noted here that the use of the stress function is to disregard the inertia terms in the equations of in-plane motions of the points in the plate. We choose the deflection function and the stress function in the separable form[4],

$$w(x,y,t) = h \ W(x,y) \ F(t), \qquad \phi(x,y,t) = h \ \Phi(x,y) \ F^2(t)$$

(7.5)

Making use of the transformations

$$\frac{\partial w}{\partial x} = w_x = \frac{dw}{dU} U_x, \quad w_{xx} = \frac{d^2 w}{dU^2} U_x^2 + \frac{dw}{dU} U_{xx}, \quad w_y = \frac{dw}{dU} u_y, \quad etc.$$

(7.6)

Eqns.(7.1) and (7.2) finally reduce to

$$D \sum_{i=1}^{4} \lambda_i \frac{d^{5-i}W}{dU^{5-i}} hF(t) = h^3 \left[\lambda_5 \frac{d^2W}{dU^2} \frac{d\Phi}{dU} + \lambda_6 \frac{d^2\Phi}{dU^2} \frac{dW}{dU} + \lambda_7 \frac{d W}{dU} \frac{d\Phi}{dU} \right] F^3(t)$$
$$+ q - \rho h^2 W \ddot{F}, \qquad \ddot{F} = \frac{d^2 F}{dt^2}$$

(7.7)

$$\sum_{i=1}^{4} \lambda_i \frac{d^{5-i}\Phi}{dU^{5-i}} = Eh \left[\lambda_8 \frac{d^2W}{dU^2} \frac{dW}{dU} + \lambda_9 \left(\frac{dW}{dU} \right)^2 \right]$$

(7.8)

in which

$\lambda_1 = \left(U_x^2 + U_y^2 \right)^2$,

$\lambda_2 = \left[6 \left(U_x^2 U_{xx} + U_y^2 U_{yy} \right) + 8 U_x U_y U_{xy} + 2 \left(U_x^2 U_{yy} + U_y^2 U_{xx} \right) \right]$

$\lambda_3 = \left[3 \left(U_{xx}^2 + U_{yy}^2 \right) + 4 U_{xy}^2 + 4 \left(U_x U_{xxx} + U_y U_{yyy} \right) + 2 U_{xx} U_{yy} \right]$,

$\lambda_4 = \left(U_{xxxx} + 2 U_{xxyy} + U_{yyyy} \right)$,

$\lambda_5 = \left(U_x^2 U_{yy} - 2 U_x U_y U_{xy} + U_y^2 U_{xx} \right) = \lambda_6, \quad \lambda_7 = \left(2 U_{yy} U_{xx} - 2 U_{xy}^2 \right)$,

$\lambda_8 = \left(2 U_x U_y \ U_{xy} - U_x^2 \ U_{yy} - U_{xx} \ U_y^2 \right)$,

$\lambda_9 = \left(U_{xy}^2 - U_{xx} U_{yy} \right)$

Since Eqns. (7.7) and (7.8) are valid for all points of the whole domain, it is clear that

$$\iint_{\Omega} \left[D \sum_{i=1}^{4} \lambda_i \frac{d^{5-i}W}{dU^{5-i}} hF(t) - h^3 \left\{ \begin{array}{l} \lambda_5 \dfrac{d^2W}{dU^2}\dfrac{d\Phi}{dU} + \lambda_6 \dfrac{d^2\Phi}{dU^2}\dfrac{dW}{dU} + \\[2mm] \lambda_7 \dfrac{dW}{dU}\dfrac{d\Phi}{dU} \end{array} \right\} F^3(t) \right. \\ \left. -q + \rho h^2 W\ddot{F} \right] d\Omega = 0$$

(7.9)

$$\iint_{\Omega} \left[\sum_{i=1}^{4} \lambda_i \frac{d^{5-i}\Phi}{dU^{5-i}} - Eh \left\{ \begin{array}{l} \lambda_8 \dfrac{d^2W}{dU^2}\dfrac{dW}{dU} + \\[2mm] \lambda_9 \left(\dfrac{dW}{dU}\right)^2 \end{array} \right\} \right] d\Omega = 0$$

(7.10)

On performing the required contour integrals, one gets

$$h F(t) \sum_{i=1}^{4} f_{1i} \frac{d^{4-i}W}{dU^{4-i}} + \frac{h^3}{2D} f_{15} \frac{dW}{dU}\frac{d\Phi}{dU} F^3(t) + \frac{f_{16}}{D} q + \frac{\rho h^2}{D} f_{17} \ddot{F}(t) \int_{u_*}^{u} WdU \Bigg|_0 = 0$$

(7.11)

$$\sum_{i=1}^{4} g_{1i} \frac{d^{4-i}\Phi}{dU^{4-i}} + \frac{Eh}{2} g_{15} \left(\frac{dW}{dU}\right)^2 = 0$$

(7.12)

here $U^* = U\,(max)$ (Fig. 1) and f_i and g_i are functions of U and U^* only. For our present study we will initially investigate the case of an annular plate, outer boundary being clamped and the inner one being free, vibrating at moderately large amplitude with an aim to use the method for plates having a mixed boundary condition.

To be brief we would like to refer to the method explained earlier in Section 6 and write only the final equation with respect to an annular plate having its outer boundary clamped and a free inner boundary.

The equation of the iso-deflection curve in this case can reasonably be taken as
$U = 1 - (x^2 + y^2)/a^2$ when the contour C_0 $U = U^* = 1 - (x^2 + y^2)/a^2 = 1 - b^2/a^2$ (Fig. 3)

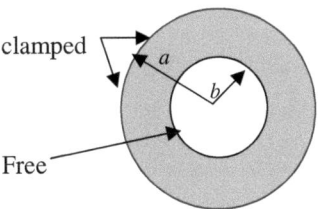

clamped

a

b

Free

Figure 3 Defining the contour

The problem has already been studied[13] so we will not discuss it in detail. But mention must be made with regard to proper use of polynomial expression for the deflection function. We often use polynomial expressions for deflection function as an approximate solution. It has been observed that there might be several expressions which satisfy the prescribed boundary conditions but yet we are to choose the proper one carefully that leads to near perfection. For example, W was assumed successively in the form as,

(i) $W = A\,u^2 + B\,u^3$
(ii) $W = A\,u^2 + B\,u^4$
(iii) $W = A\,u^2 + B\,u^3 + C\,u^4,$

all of which satisfy the required boundary conditions. For $v = 0.3$, the approximate values for W found to be
(i) $W = u^2 + 16.8889\,u^3$
(ii) $W = u^2 + 0.8042\,u^4$
(iii) $W = u^2 - 1.37766\,u^3 + 0.8698\,u^4$

However, the third one appeared to be the best of the three. Of all approximate methods it has been observed that a Galerkin method in combination with a polynomial expression for the deflection function yields very satisfactory results. The following supports the above argument (Table 2, Table 3).

Table 2 Frequency parameter $\Omega^* = \omega a^2 \sqrt{\dfrac{\rho h}{D}}$ for a clamped annular plate $\nu =$ 0.3, (b/a = 0.5)

	Finite Strip Ref.16	FEM Ref.23	REF. [1]	Present
$\Omega^* =$	17.747	17.85	17.70	17.693

Table 3 Comparison of results for dynamic and static cases

Dynamic Response [T*/T]			Static Deflection		
	Rigid circular plate (RCP)	Annular Plate (AP)	$q a^4 / E h^4$		
Relative Amplitude (ξ)	Present & Ref. [4]	Present	W_{max}/h	(RCP)	(AP)
0.25	0.9891	0.9459	0.25	1.3084	5.1063
0.50	0.9585	0.8246	0.50	3.2758	14.3680
0.75	0.9133	0.6969	0.75	5.5610	31.9400
1.00	0.8596	0.5890	1.00	8.6230	61.9787
				8.6020 [4]	
				9.0000 [23]	
1.25	0.8026	0.5037	1.25	12.7208	108.6385
1.50	0.7463	0.4370	1.50	18.1133	176.0750
1.75	0.6930	0.3844	1.75	25.0593	268.4436
2.00	0.6437	0.3424	2.00	33.8180	389.9000

7.2 Large Amplitude Vibration of a shallow Dome upon an Elliptic Base

Consider the vibration of a shallow dome of nonzero Gaussian curvature upon an elliptic base. Fig. 2 and 3 depicts the geometry of the shell. The edges are clamped and immovable. When the shell vibrates in a normal mode, the lines of equal deflections, as described in Sec.4, may reasonably be taken as

$$U(x, y) = 1 - \frac{x^2}{a^2} - \frac{y^2}{b^2}$$

(7.13)

Clearly, in this case $U = 0$ on the boundary and $U = U^* = 1$ at the centre of the shell.

The corresponding values of $\lambda_i \ (i = 1,2,......,12)$ are given by

$$\lambda_1 = 16\left(\frac{x^2}{a^4} + \frac{y^2}{b^4}\right)^2, \qquad \lambda_2 = -48\left(\frac{x^2}{a^6} + \frac{y^2}{b^6}\right), \qquad \lambda_3 = 4\frac{3a^4 + 2a^2b^2 + 3b^4}{a^2b^2}$$

$$\lambda_4 = 0, \qquad \lambda_5 = -\frac{8(1-u)}{a^2b^2} = \lambda_9, \qquad \lambda_6 = \frac{8}{a^2b^2} = \lambda_{10}$$

$$\lambda_7 = 4\left(\frac{x^2}{a^4 R_y} + \frac{y^2}{b^4 R_x}\right) = \lambda_{11}, \qquad \lambda_8 = -2\left(\frac{1}{a^2 R_y} + \frac{1}{b^2 R_x}\right) = \lambda_{12}$$

(7.14)

Substituting the above values in Eqns.(5.5) and (5.6), and utilizing the formula given by (5.7) one gets

$$\left[(1-U)^2 \frac{d^4 w}{dU^4} - 4(1-U)\frac{d^3 w}{dU^3} + 2\frac{d^2 w}{dU^2}\right] = -\alpha\,[(1-U)((1-u)\frac{dw}{dU}\frac{dF}{dU}\ )$$

$$- \beta\,\frac{d}{dU}\left[(1-U)\frac{dF}{dU}\right] + \frac{q}{2DP} - \frac{\rho\,h}{2DP}w_{,tt}$$

(7.15)

$$\left[(1-U)^2 \frac{d^4 F}{dU^4} - 4(1-U)\frac{d^3 F}{dU^3} + 2\frac{d^2 F}{dU^2} \right] = \gamma \left[\frac{d}{dU}\{(1-U)(\frac{dw}{dU})^2 \} \right]$$

$$+ \delta \frac{d}{dU}\left[(1-U)\frac{dw}{dU} \right]$$

(7.16)

where

$$\alpha = \frac{4h}{DPa^2 b^2}, \qquad \beta = \frac{h\,\kappa}{DP} \qquad\qquad \gamma = \frac{2E}{Pa^2 b^2},$$

$$\delta = \frac{E\,\kappa}{P} \qquad P = \frac{3a^4 + 2a^2 b^2 + 3b^4}{a^4 b^4} \qquad \kappa = \left(\frac{1}{a^2 R_y} + \frac{1}{b^2 R_x} \right)$$

(7.17)

It appears that the exact solution of Eqns. (7.15) and (7.16) is not possible to find. So in order to obtain an approximate solutions, let us assume

$$w(U,t) = W(U)\, f(t) \approx AU^2\, f(t)$$

(7.18)

where f(t) is an unknown function of time to be determined.

A question may arise as to the validity of using a single term in the above approximation. It has been observed that considering more number of terms in eqn (7.18), the amount of mathematical computation is enormous whereas the success is negligible. This happens for structures having simple boundaries, whereasfor non-simple boundaries or with mixed boundary conditions one should take care and approximation should be made wisely. We will discuss it in detail later on.

Substitution of Eqn. (7.18) in Eqn. (7.16) and taking the first integral of (7.16) it yields

$$\frac{d}{dU}\left\{ (1-U)\frac{dF}{dU} \right\} = \frac{4\gamma}{3} A^2 f^2(t)\, U^3 + \delta A f(t)\, U^2 + A_1$$

(7.19)

which on further integration reduces to

$$\left\{(1-U)\frac{dF}{dU}\right\} = \frac{\gamma}{3}A^2 f^2(t)U^4 + \frac{1}{3}\delta A f(t)U^3 + A_1 U + A_2$$

$$(7.20)$$

Considering the case for a *clamped immovable edge* condition in which we set the following conditions as explained in Ref. [9]

$$\frac{dF}{dU}\bigg|_{U=0} = 0 \quad \text{and} \quad \left\{(1-U)\frac{d^2F}{dU^2} - 2(1-v)\frac{dF}{dU}\right\}\bigg|_{U=0} = 0 \ (\text{Ref. 9})$$

$$(7.21)$$

This gives both A_1 and A_2 to be zero and Eqn. (7.20) reduces to

$$\left\{(1-U)\frac{dF}{dU}\right\} = \frac{\gamma}{3}h^2 f^2(t)U^4 + \frac{1}{3}\delta h f(t)U^3$$

$$(7.22)$$

Eqn. (7.22), after applying Galerkin's method as in the foregoing analysis, yields a Duffing-type equation in f(t):

$$\rho h^2 \ddot{f} + \alpha_1 f + \alpha_2 f^2 + \alpha_3 f^3 = Q*$$

$$(7.23)$$

where

$$\alpha_1 = \frac{DP}{3}\left(\frac{40}{3} + 2\beta \ \delta\right)A \qquad \alpha_2 = \frac{10DP}{9}\left(\frac{2}{9}\beta\gamma + \frac{4}{9}\delta\alpha\right)A^2$$

$$\alpha_3 = \frac{100 \ DP\gamma\alpha \ A^3}{21}, \qquad Q* = \frac{5}{3}\frac{q}{3}$$

$$(7.24)$$

and

$$\alpha = \frac{4h}{DPa^2b^2}, \qquad \beta = \frac{h}{DP}\left(\frac{1}{a^2 R_y} + \frac{1}{b^2 R_x}\right) \qquad \gamma = \frac{2E}{Pa^2b^2},$$

$$\delta = \frac{E}{P}\left(\frac{1}{a^2 R_y} + \frac{1}{b^2 R_x}\right) \qquad P = \frac{3a^4 + 2a^2b^2 + 3b^4}{a^4 b^4}$$

$$(7.25)$$

An indirect verification of the correctness of the time differential equation may be made by considering the case for a flat plate. When $\beta \to 0$, $\delta \to 0$, it implies that $\alpha_2 = 0$ and further if $a = b$ the problem reduces to that of a circular plate for which Eqn. (7.23) takes the form (for $v = 0.3$)

$$\rho h^2 \ddot{f} + \frac{Eh^4}{a^4}\left[9.756\, f + 4.762\, f^3\right] = \frac{5}{3}q \qquad (present\ study)$$

whereas

$$\rho h^2 \ddot{f} + \frac{Eh^4}{a^4}\left[9.768\, f + 4.602\, f^3\right] = \frac{5}{3}q \qquad (Re\ f.4)$$

$$(7.26)$$

Which is in excellent agreement considering the fact that only a single term approximation for the deflection function has been considered for the present study.

7.2(a)   Linear Free Vibration

If we set, α_2, α_3 and $Q*$ each equals to zero, then the linear frequency is given by

$$\omega_L^2 = \frac{DP}{\rho h}\left[\frac{40}{3} + 2\beta\delta\right] \quad or \quad \omega_L^2 = \frac{Eh^2 P}{8\rho}\left[\frac{(320/3)}{12(1-v^2)} + 4\left(\frac{2\gamma_1}{h}\right)^2\right]$$

$$(7.27)$$

where $\gamma_1 = \kappa/P$ and $\dfrac{2\gamma_1}{h} = \gamma*$ represents the measure of shallowness of the shell.

Eq. (7.27), on simplification and with a little rearrangement of the parameters, becomes

$$\omega_L^2 = \frac{Eh^2 P}{8\rho}\left[\frac{\lambda^4}{12(1-v^2)} + \frac{M^4}{12(1+v)^2}\right]$$

(7.28)

where $M^4 = \frac{192\gamma_1^2}{h^2}(1+v)^2$ and $\lambda^4 = (320/3) \approx (3.196)^4$ (for fundamental mode

of vibration) has been introduced for comparison of results given in Ref. [10]. If ω_0 be the value of ω_L corresponding to $M=0$ and $v=0$, that is the value of the frequency for a flat plate with vanishing Poisson's ratio, then

$$\omega_0 = 2.984\left[Eh^2(3a^4 + 2a^2b^2 + 3b^4)/(8\rho a^4 b^4)\right]^{1/2}$$

(7.29)

and

$$\frac{\omega_L}{\omega_0} = \left[\left(\frac{1}{(1-v^2)}\right) + \left(\frac{M}{\lambda_0}\right)^4 \frac{1}{(1+v)^2}\right]^{1/2}$$

(7.30)

which are in exact agreement with that of Ref.[10].
If the second term in the above expression for ω_L dominates the first when M is large, then

$$\omega_{L2} = \left(\frac{2E}{P\rho}\kappa^2\right)^{1/2},$$

Which is exactly the same as that of Ref.[10]. It may be noted here that following Reissner[24] the first term is predominant when γ^* or $H/h < or =$ 25 and the second term is predominant when $H/h \geq 25$ in order that the theory of shallow shells is applicable. Table 4 shows a close agreement for the values of fundamental frequency for a flat circular plate.

Table 4 Values of coefficient of in the expression for the fundamental frequency for a circular plate.

v	Ref. 24	Present Study	Ref.4
0	2.948	2.948	____
0.3	3.091	3.125	3.125

7.2(b)   Nonlinear Free Vibration

Substituting $Q^* = 0$ in Eqn. (7.23) one obtains

$$\rho h^2 \ddot{f} + \alpha_1 f + \alpha_2 f^2 + \alpha_3 f^3 = 0$$

(7.31a)

or

$$\ddot{f} + A_1 f + A_2 f^2 A_3 f^3 = 0$$

(7.31b)

This is a familiar form of time differential equation and for which the frequency ratio (Nonlinear to Linear) is given by[16],

$$\frac{\omega^*}{\omega} = \left[1 + \left\{ \frac{3}{4} \frac{A_3}{A_1} - \frac{5}{6} \left(\frac{A_2}{A_1} \right)^2 \left(\frac{A}{h} \right)^2 \right\} \right]^{1/2}.$$

(7.32)

from which one can find the nonlinear effect on the frequency. The results have been presented in the form of graphs (Figs. 8 - 10) and Columns 1 and 2 of Table 5 below.

Table 5 Nonlinear effect on the frequency by varying values of (ω^*/ω)

	Values of (ω^*/ω) for a/b=1 for different values of v and γ^*			Values of (ω^*/ω) for a/b=1.5 and γ^*=5 for different values of v		
	$v = 0$	$v = 0.3$	$v = 0.5$	$v = 0$	$v = 0.3$	$v = 0.5$
γ^*	(ω^*/ω)	(ω^*/ω)	(ω^*/ω)			
0	1.077697	1.070936	1.058808	1	1	1
0.5	1.054078	1.048371	1.038602	0.996857	0.996912	0.997036
1	1.495356	1.452167	1.374265	0.987368	0.987589	0.988091
1.5	2.13267	2.04855	1.892266	0.971348	0.971853	0.973
2	2.817702	2.696798	2.469124	0.948465	0.949384	0.951471
2.5	3.512205	3.357009	3.062966	0.918206	0.919689	0.923053
3	4.207663	4.019331	3.661462	0.879812	0.882039	0.887084
3.5	4.902489	4.681548	4.261079	0.832155	0.835358	0.842598
5	6.983613	6.665906	6.06027	0.773527	0.778023	0.788154
10	13.91613	13.27773	12.05965	0.70118	0.707448	0.721502
20	27.79355	26.51485	24.07448	0.610256	0.619123	0.638831

7.2(c)   Static Deflection

Neglecting the inertial term, Eqn.(7.23) can be written as

$$\alpha_1 f + \alpha_2 f^2 \alpha_3 f^3 = \frac{5}{3} q$$

(7.33)

which after simplification reduces to (here f stands for maximum static deflection).

$$\frac{qa^4}{Eh^4} = L\left[\left\{\frac{2}{3(1-v^2)} + 0.3\left(\frac{2\gamma_1}{h}\right)^2\right\} f + \frac{20}{3N}\left(\frac{2\gamma_1}{h}\right) f^2 + \frac{160}{7N^2} f^3\right]$$

(7.34)

where

$$L = \left\{3\left(\frac{a}{b}\right)^4 + 2\left(\frac{a}{b}\right)^2 + 3\right\}, \quad N = \left\{3\left(\frac{a}{b}\right)^2 + 2 + 3\left(\frac{b}{a}\right)^2\right\}$$

(7.35)

Since in the literature, no result on the static large deflection of a dome on an elliptic base is available, we may verify the results with that for a flat circular plate in the limiting case. When $\kappa \to 0$, $a = b$, equation (7.33) represents the static behaviour of a flat circular plate of radius a with clamped immovable edges. Equation (7.34) shows a comparative study.

$$\frac{qa^4}{Eh^4} = \begin{cases} 5.8608\,\overline{w}_m + 2.857\,\overline{w}_m^3 & (present\ study) \\ 5.861\overline{w}_m + 2.761\,\overline{w}_m^3 & (Ref.4\ ) \\ 5.848\,\overline{w}_m + 2.754\,\overline{w}_m^3 & (.Ref\,25\ ) \end{cases}$$

(7.36)

where $\overline{w}_m$ stands for

$$\overline{w}_m = w_m / h = [f(t)]_{max} \quad [\text{Ref. 4}]$$ (7.37)
$$= (\text{maximum deflection divided by the plate thickness})$$

The graphical representation of the above results has been made in Fig. 4 validating the correctness of the present method.

7.2(d)   Results and Discussion

Frequency Analysis:

Table 2 shows the values of linear frequency for a circular plate obtained using different approaches. It justifies the present approach (CDC method). Further discussion on the linear frequency is considered to be irrelevant as above are exactly the same as those obtained in Refs. [9, 24] and the authors have already made detailed discussions on it.

Static Analysis:

The results for a shallow shell resting on an elliptical base have been shown in figures 4-7. Fig.4 gives a comparison of results for maximum deflection for a circular plate obtained through a classical method and through the CDC-method.

Fig.5 shows the load-deflection behaviour for a spherical shell for different values of v and $\gamma^* = (2\gamma/h)$. It shows that there is no significant difference for the load-deflection curve for a spherical shell for $\gamma^* = (2\gamma/h) < 5$. But the measure of shallowness affects the results when $\gamma^* = (2\gamma/h) \geq 5$ and greater is the measure, lower is the deflection. Fig. 5 shows the effect of $\gamma^* = (2\gamma/h)$ on the load-deflection curve of the shell for a fixed ratio of the aspect ratio of the elliptic base. In this case it is observed that greater is the measure of shallowness lower is the deflection. Fig.5 shows that for a particular load, deflection increases with the increase of shallowness of the shell.

A comparison of results shown in Figs. 6 and 7 indicate that for a certain load the deflection increases with the increase of $\gamma^* = (2\gamma/h)$ or with the increase in the aspect ratio (a/b).

Vibration Analysis:

Figs. 8-15 show the dependence of nonlinear to linear frequency ratio on γ^* and aspect ratio a/b. Figs. 8 and 11 show the result for a spherical shell ($\gamma^*=0$) . It has been observed that the dependence on the Poisson's ratio is not so much significant though the nonlinear effect is comparatively a little lower for higher values of v. Figs. 14 makes a comparative study of dependence of the relative frequency ratio γ^* on the aspect ratio (a/b) of the axes of the elliptic base of the dome. The nonlinear effect is significant when the value of a/b decreases. Fig. 14 confirms that the nonlinear effect is not so much dependent on aspect ratio for $\gamma^* \geq 1.5$ - 2. Considering all aspects as relevant from the Figs. 8-15, it appears that the values of γ^* in the range of 1 – 2 affect the nonlinear behaviour of the vibrating shell.

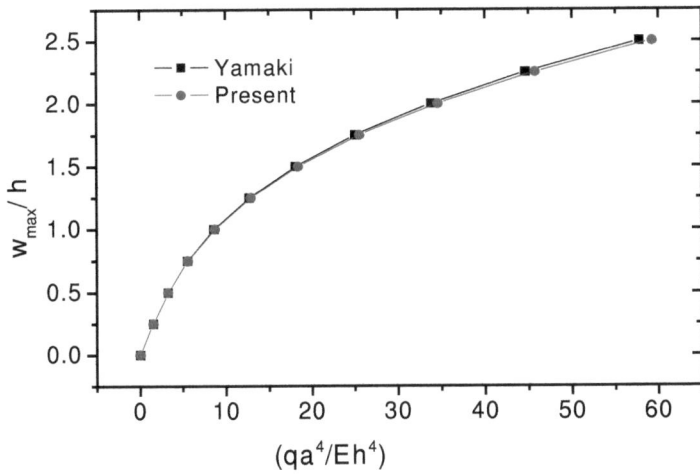

Figure 4 Comparison of result with that of Yamaki [4]

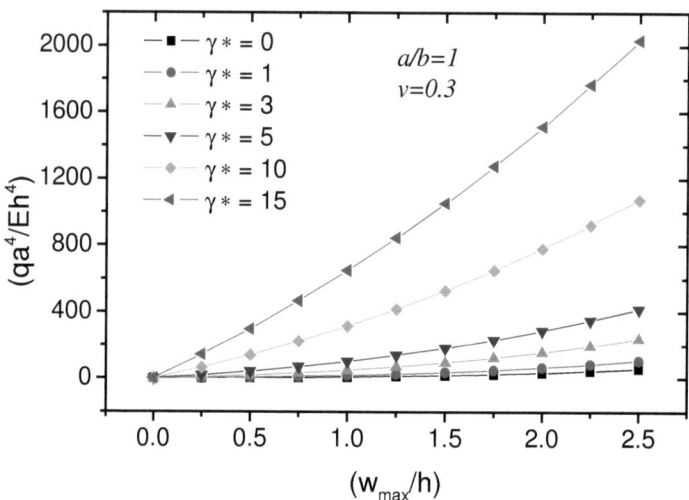

Figure 5 Load-Deflection Curve for a spherical shell of radius a, and 0, 1, 3, 5, 10 and 15.

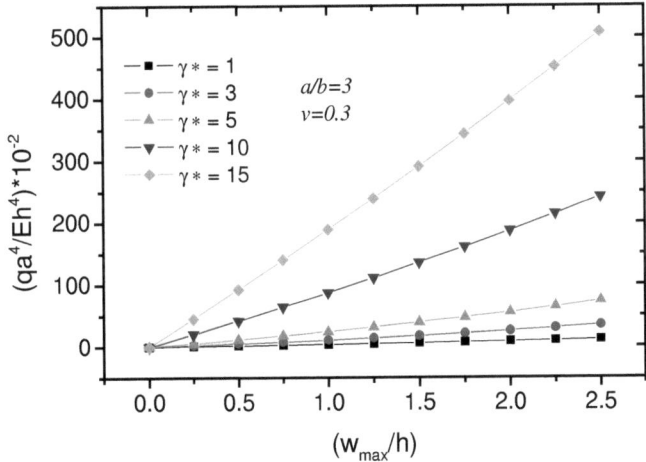

Figure 6 Load-Deflection Curve for (a/b) = 3,   and =1, 3, 5, 10 and 15.

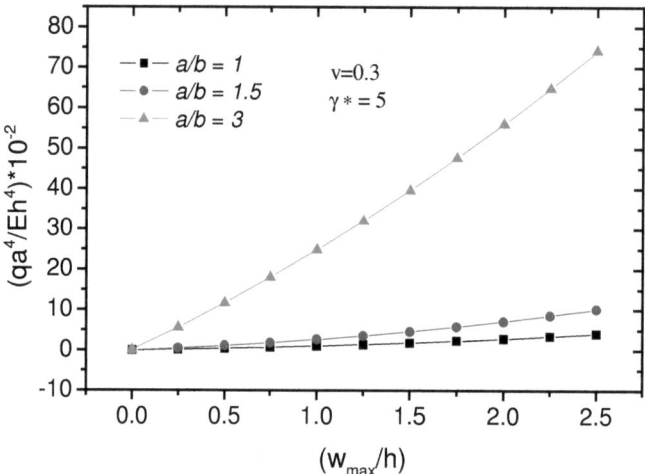

Figure 7 Load-Deflection curves for a/b = 1, a/b = 1.5, a/b = 3 with $\gamma* = 5$

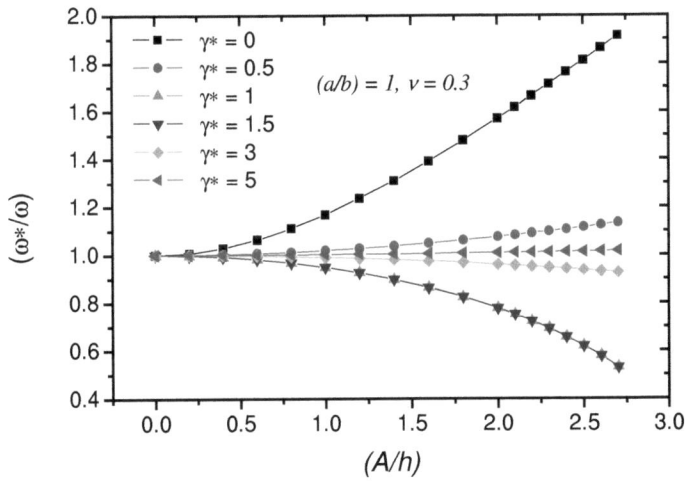

Figure 8 (ω^*/ω) vs. relative amplitude (A/h) for a spherical shell.

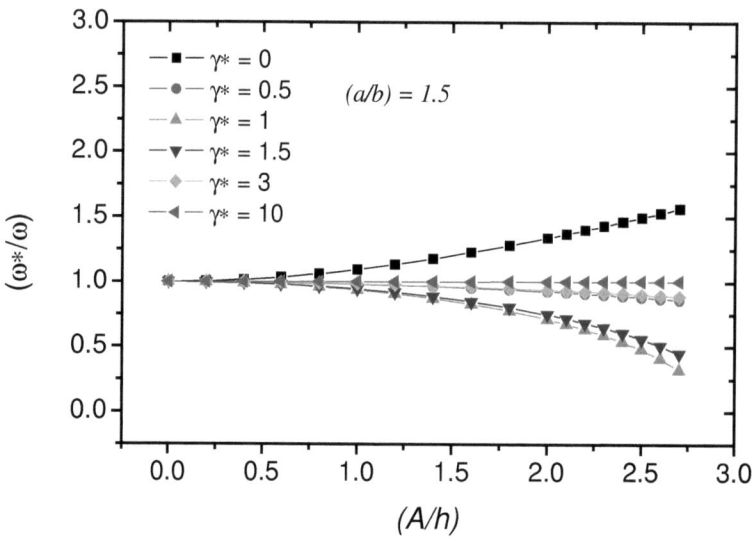

Figure 9 $(\omega*/\omega)$ vs. relative amplitude (A/h) for a/b =1.5, v=0.3

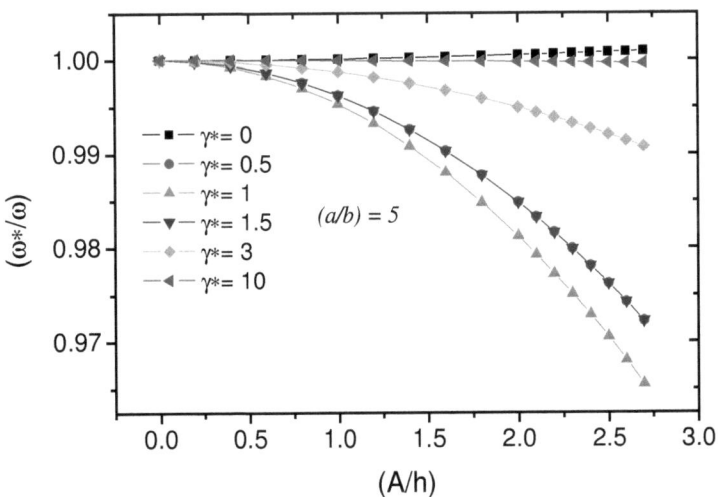

Figure 10 (ω^*/ω) vs. relative amplitude (A/h) for a/b= 5, v=0.3.

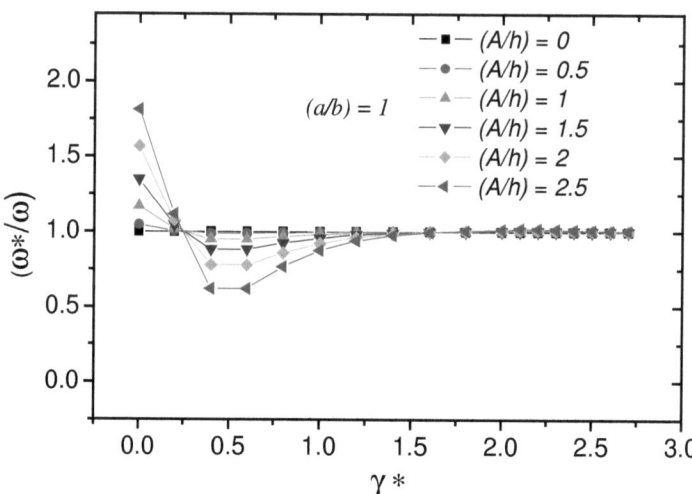

Figure 11 (ω^*/ω) vs. measure of shallowness for a spherical shell for various amplitudes.

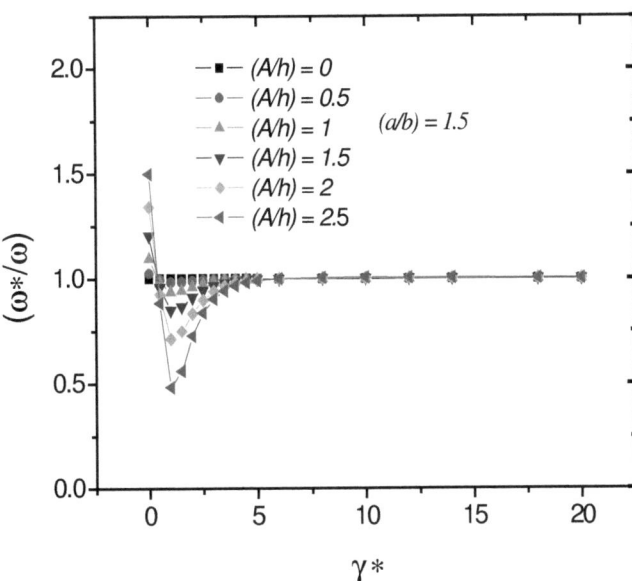

Figure 12 (ω^*/ω) vs. measure of shallowness for aspect ratio 1.5 and for various amplitudes.

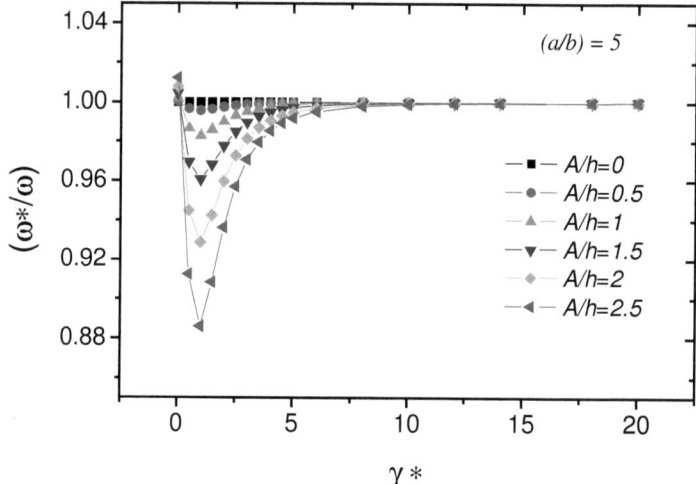

Figure 13 (ω^*/ω) vs. measure of shallowness for various amplitudes for aspect ratio a/b = 5

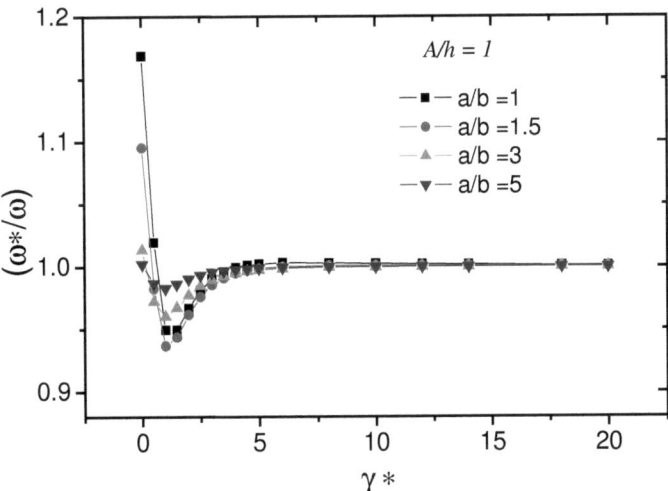

Figure 14 (ω^*/ω) vs. measure of shallowness for various aspect ratio for relative amplitude $A/h = 1$.

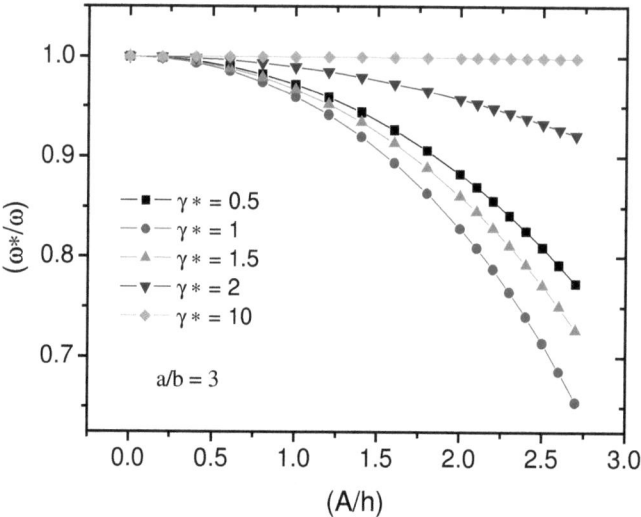

Figure 15 (ω^*/ω) vs. relative amplitude A/h for various values of measure of shallowness and for aspect ratio a/b = 3.

7.2(e)   A Remark on CDC Method

In conclusion it can be said that the method proposed in this study offers a new approach to deal with problems involving large amplitude vibrations of plates and shallow shells. The application of polynomial expressions for the deflection and the stress functions in conjunction with the Galerkin procedure appears to produce highly accurate results. The comparison of results shows that using a moderately approximated expression for the deflection function yields results, which are comparable to the previously obtained results using other approximate methods. It can therefore be concluded that the CDC method appears to be a simple tool to deal with the problems of nonlinear vibration of plates and shallow shells of arbitrary shapes.

Chapter 8
SCOPE FOR FURTHER RESEARCHES ON THE APPLICATION OFTHE CDC METHOD TO ELASTIC-PLASTIC SHALLOW SHELLS

The present chapter includes briefly the nonlinear vibration analysis of elastic-plastic shallow shells of arbitrary shapes. The basic governing equations for nonlinear vibration analysis for such shells have been deduced. The authors invite all interested researchers to develop the present work in future to help the designers in the field of civil, mechanical, offshore and aeronautical structures.

In modern technology the plates and shells are often encounter very high transverse loads and consequently the plates or shells become quite difficult to analyse when they have arbitrary geometries. Hence the situation necessitates to find a simple, reliable and that could provide accurate results. The most commonly used, the finite element technique which is used in most cases. But there appear to exist certain classes of problems in which properties of that class may be exploited in the formulation with a view to reduce the degrees of freedom, hence requiring less storage and execution time. Mzumdar and Jain[28] presented the pertinent equations for a plate of arbitrary shape. The CDC method is then applied to illustrate the elastic-plastic bending of simply supported elliptic plates.

Here the authors intend to establish the governing equations of shallow shells vibrating at moderately large amplitudes. A simple illustration is also made to compare the results with the existing literature. The present study may, therefore, be regarded as a sequel to earlier study to deal with the nonlinear vibration of shallow shells based on the CDC Method.

8.1   Derivation of Governing Equations and the Time Differential Equation

Following Ilyushin's approach to Elastic-Plastic deformation [26] the bending moments M_x, M_y and M_{xy} are given by the following relationships

$$M_x = -D_0(1-\Omega)[w_{xx} + v\,w_{yy}], \quad M_y = -D_0(1-\Omega)[w_{,yy} + v\,w_{,xx}]$$

and
$$M_{xy} = -D_0(1-\Omega)(1-v)\,w,_{xy}$$

(8.1)

Furthermore, since D and E are related as $D = Eh^3/12(1-v^2)$, when h and v are constants, D/E *is a constant*. So when D assumes the form $D = D_0(1-\Omega)$, E changes to $E = E_0(1-\Omega)$

where
$$\Omega = \lambda\left(1 - \frac{3}{2e} + \frac{1}{2e^3}\right).$$

(8.2)

Here $\Omega = 0$, when $e \leq 1$ and the region is elastic and the region is plastic when $e \geq 1$,

and
$$e^2 = \frac{h^2}{3e_s}\left(w_{xx}^2 + w_{yy}^2 + w_{xx}w_{yy} + w_{xy}^2\right)$$

(8.3)

in which e_s is the yield strain and λ is a material constant. With usual notations using the expressions for the total strain energy, the kinetic energy and the work done and then formulating the Lagrangian and applying Hamilton's principle to it, a straightforward application of the variational calculus yields the following equations of motion[3]

$$(1-\Omega)\nabla^4 w - \frac{\partial\Omega}{\partial x}\left(2w_{xxx} + 2w_{xyy}\right) - \frac{\partial\Omega}{\partial y}\left(2w_{yyy} + 2w_{xxy}\right) - \frac{\partial^2\Omega}{\partial x^2}\left(w_{xx} + v\,w_{yy}\right) - \frac{\partial^2\Omega}{\partial y^2}\left(w_{yy} + v\,w_{xx}\right)$$

$$-2(1-v)\frac{\partial^2\Omega}{\partial x\partial y}w_{xy} = \frac{h}{D_0}S(w,F) - \frac{h}{D_0}\left(\frac{F_{yy}}{R_x} + \frac{F_{xx}}{R_y} - 2\frac{F_{xy}}{R_{xy}}\right) - \rho\frac{h}{D_0}\ddot{w} + \frac{q}{D_0}$$

(8.4)

and

$$\nabla^4 F = E_0(1-\Omega)[\left(w^2_{xy} - w_{xx}w_{yy}\right) + \left(\frac{w_{yy}}{R_x} + \frac{w_{xx}}{R_y} - 2\frac{w_{xy}}{R_{xy}}\right)]$$

$$- E_0\frac{\partial\Omega}{\partial x}\left(w\,w_{xy} - w_x w_{yy} + 2\frac{w_x}{R_y} - 2\frac{w_y}{R_{xy}}\right) -$$

$$- E_0\frac{\partial\Omega}{\partial y}\left(w_x w_{xy} - w_{xx}w_y + 2\frac{w_y}{R_x}\right) - E_0\frac{\partial^2\Omega}{\partial x^2}\left(\frac{1}{2}w^2_y + \frac{w}{R_y}\right) - E_0\frac{\partial^2\Omega}{\partial y^2}\left(\frac{1}{2}w^2_x + \frac{w}{R_x}\right)$$

$$+ E_0\frac{\partial^2\Omega}{\partial x\partial y}(w_x w_y + 2\frac{w}{R_{xy}})$$

(8.5)

where suffixes in 'w' and 'F' denote partial differentiation w.r.t. the variables.

$$S(w,F) \equiv \frac{\partial^2 w}{\partial x^2}\frac{\partial^2 F}{\partial y^2} - 2\frac{\partial^2 w}{\partial x \partial y}\frac{\partial^2 F}{\partial x \partial y} + \frac{\partial^2 w}{\partial y^2}\frac{\partial^2 F}{\partial x^2} \cong w_{,xx}F_{,yy} - 2w_{,xy}F_{,xy} + w_{,yy}F_{,xx},$$

Here 'F' denotes well-known Airy-Stress function.

Let $U = U(x,y) = constant$ be a member of the family of iso-deflection or iso-amplitude contour lines. Using the following transformations as before

$$\frac{\partial w}{\partial x} = w_x = \frac{dw}{dU}U_x \ , \ \frac{\partial w}{\partial y} = w_y = \frac{dw}{dU}U_y \ ,$$

$$w_{xx} = \frac{d^2 w}{dU^2}U_x^2 + \frac{dw}{dU}U_{xx} \ w_{xy} = \frac{d^2 w}{dU^2}U_x U_y + \frac{dw}{dU}U_{xy}, \text{etc}\ldots$$

Equations (8.4, 8.5) can be transformed into the following

$$\sum_{i=1}^{4} \lambda_i \frac{d^{5-i} w}{du^{5-i}} - \frac{d\Omega}{dU}\left[2\lambda_5 \frac{d^3 w}{dU^3} + \lambda_6 \frac{d^2 w}{dU^2} + \lambda_7 \frac{dw}{dU}\right]$$

$$- \frac{d^2 \Omega}{dU^2}\left[\lambda_8 \frac{d^2 w}{dU^2} + \lambda_9 \frac{dw}{dU}\right]$$

$$= \frac{h}{D_0}\left[\lambda_{10}\frac{d}{dU}\left(\frac{dw}{dU}\frac{dF}{dU}\right) + \lambda_{11}\frac{dw}{dU}\frac{dF}{dU}\right]$$

$$- \frac{h}{D_0}\left[\lambda_{12}\frac{d^2 F}{dU^2} + \lambda_{13}\frac{dF}{dU}\right] \qquad -\frac{q}{D_0} + \frac{\rho h}{D_0}w_{,tt}$$

$$(8.6)$$

$$\sum_{i=1}^{4} \Lambda_i \frac{d^{5-i}F}{dU^{5-i}} = E_0 \left(1 - \Omega\right) \left[\begin{array}{c} \Lambda_{14} \dfrac{d^2 w}{dU^2} \dfrac{dw}{dU} + \Lambda_{15} \left(\dfrac{dw}{dU}\right)^2 + \Lambda_{16} \dfrac{d^2 w}{dU^2} \\[2mm] + \Lambda_{17} \dfrac{dw}{dU} \end{array} \right]$$

$$- E_0 \frac{d\Omega}{dU} \left[\frac{1}{2} \Lambda_{14} \left(\frac{dw}{dU}\right)^2 + 2\Lambda_{16} \frac{dw}{dU} + \Lambda_{17} w \right]$$

$$- E_0 \frac{d^2\Omega}{dU^2} \left[\Lambda_{16} w \right]$$

$$(8.7)$$

where

$$\Lambda_1 = \left(U_x^2 + U_y^2\right)^2 \qquad \Lambda_2 = U_x^2\left(6U_{xx} + 2U_{yy}\right) + U_y^2\left(6U_{yy} + 2U_{xx}\right) + 8U_x U_y U_{xy}$$

$$\Lambda_3 = 3\left(U_{xx}^2 + U_{yy}^2\right) + 4\left(U_x U_{xxx} + U_y U_{yyy}\right) + \left(U_x U_{xyy} + U_y U_{xxy}\right) + 4U_{xy}^2 + 2U_{xx} U_{yy}$$

$$\Lambda_4 = \left(U_{xxxx} + U_{yyyy} + 2U_{xxyy}\right),$$

$$\Lambda_5 = \left(U_x^2 + U_y^2\right)^2 = \Lambda_1 = \Lambda_8 \ ,$$

$$\Lambda_6 = 6\left(U_x^2 U_{xx} + U_y^2 U_{yy}\right) + 2\left(U_x^2 U_{yy} + U_y^2 U_{xx} + 8U_x U_y U_{xy}\right) +$$
$$\left(U_x^2 U_{xx} + U_y^2 U_{yy} + 2U_x U_y U_{xy}\right) + 2\nu\left(U_x^2 U_y^2 - U_x U_y U_{xy}\right)'$$

$$\Lambda_7 = 2(U_x U_{xxx} + U_y U_{yyy} + U_x U_{xyy} + U_y U_{xxy}) + U_{xx}^2 + U_{yy}^2 + 2U_x U_y U_{xy} + 2\nu\left(U_{xx} U_{yy} - U_{xy}^2\right),$$

$$\Lambda_9 = \left(U_x^2 U_{xx} + U_y^2 U_{yy} + 2U_x U_y U_{xy}\right) + \nu\left(U_x^2 U_{yy} + U_y^2 U_{xx} - 2U_x U_y U_{xy}\right),$$

$$\Lambda_{10} = \left(U_x^2 U_{yy} + U_y^2 U_{xx} - 2U_x U_y U_{xy}\right) = -\Lambda_{14},$$

$$\Lambda_{11} = 2\left(U_{xx} U_{yy} - U_{xy}^2\right) = -2\Lambda_{15}$$

$$\Lambda_{12} = \left(\frac{U_y^2}{R_x} + \frac{U_x^2}{R_y} - 2\frac{U_x U_y}{R_{xy}}\right) = \Lambda_{16}$$

$$\Lambda_{13} = \left(\frac{U_{yy}}{R_x} + \frac{U_{xx}}{R_y} - 2\frac{U_{xy}}{R_{xy}}\right) = \Lambda_{17}$$

$$(8.8)$$

are all functions of partial derivatives of U

Since Eqns. (8.4) and (8.5) are valid for all points on the surface of the shell, we can have from Eqns. (8.6) and (8.7)

$$\iint\limits_{A} (\sum_{i=1}^{4} \lambda_i \frac{d^{5-i} w}{dU^{5-i}} - \frac{d\Omega}{dU} \left[2\lambda_5 \frac{d^3 w}{dU^3} + \lambda_6 \frac{d^2 w}{dU^2} + \lambda_7 \frac{dw}{dU} \right] -$$

$$\frac{d^2\Omega}{dU^2} \left[\lambda_8 \frac{d^2 w}{dU^2} + \lambda_9 \frac{dw}{dU} \right] - \frac{h}{D_0} \left[\lambda_{10} \frac{d}{dU} \left(\frac{dw}{dU} \frac{dF}{dU} \right) + \lambda_{11} \frac{dw}{dU} \frac{dF}{dU} \right]$$

$$+ \frac{h}{D_0} \left[\lambda_{12} \frac{d^2 F}{dU^2} + \lambda_{13} \frac{dF}{dU} \right] + \frac{q}{D_0} - \frac{\rho h}{D_0} w,_{tt}) dA = 0 \qquad (8.9)$$

and

$$\iint\limits_{A} \sum_{i=1}^{4} \Lambda_i \frac{d^{5-i} F}{dU^{5-i}} dA$$

$$= \iint\limits_{A} \{ E_0 (1 - \Omega) \left[\Lambda_{14} \frac{d^2 w}{dU^2} \frac{dw}{dU} + \Lambda_{15} \left(\frac{dw}{dU} \right)^2 + \Lambda_{16} \frac{d^2 w}{dU^2} + \Lambda_{17} \frac{dw}{dU} \right]$$

$$- E_0 \frac{d\Omega}{dU} [\frac{1}{2} \Lambda_{14} \left(\frac{dw}{dU} \right)^2 + 2\Lambda_{16} \frac{dw}{dU} + \Lambda_{17} w] - E_0 \frac{d^2\Omega}{dU^2} [\Lambda_{16} w] \} dA$$

$$(8.10)$$

where the integration is taken over the region bounded by any contour C_U. While performing the above integrals it would be more convenient to utilize the formula in the modified form[13] and care should be taken to evaluate first the contour integral.

The method of solution has been explained in Ref. [13] to obtain the *time differential equation,* which can be represented as

$$\ddot{\Psi}(t) + L\Psi(t) + M\Psi^2(t) + N\Psi^3(t) = q*, \qquad (8.11)$$

from which all kinds of static and dynamic analysis of the structure can be made.

8.2 An Illustration

For a specific illustration with a view to compare the present results we consider the case of large amplitude vibration of a shallow Dome upon an Elliptic Base.

Consider the vibration of an elastic-plastic shallow dome of nonzero Gaussian curvature. Fig. 2 depicts the geometry of the shell. The edges are clamped and immovable. When the shell vibrates in a normal mode, the lines of equal deflections, as described in Sec.3, may reasonably be taken as

$$U(x,y) = 1 - \frac{x^2}{a^2} - \frac{y^2}{b^2}$$ (8.12)

Performing the required integrals in equations (8.9) and (8.10)) the resulting equation will take the forms

$$(1-\Omega)\left[(1-U)^2\frac{d^3w}{dU^3} - 2(1-U)\frac{d^2w}{dU^2}\right] - \frac{d\Omega}{dU}\left[(1-U^2)\frac{d^2w}{dU^2} - 2\frac{P^*}{P}\frac{dw}{dU}\right] -$$

$$= -\alpha(1-U)\left(\frac{dw}{dU}\frac{dF}{dU}\right) - \beta(1-U)\frac{dF}{dU}$$

$$-\frac{q}{2D_0P}(1-U) - \frac{\rho h}{2D_0P}\int_1^U w_{tt}\, dU = 0$$ (8.13)

and

$$(1-U)^2\frac{d^3F}{dU^3} - 2(1-U)\frac{d^2F}{dU^2}$$

$$= (1-\Omega)\left[\gamma(1-U)\frac{d^2w}{dU^2}dU + \delta(1-U)\frac{dw}{dU}\right]$$

$$-\frac{d\Omega}{dU}[\int_1^U \gamma(1-U)\left(\frac{dw}{dU}\right)^2 + \delta(1-U)w]$$

(8.14)

where $\quad \alpha = \dfrac{4h}{DPa^2b^2}, \qquad \beta = \dfrac{h}{DP}\left(\dfrac{\kappa}{a^2b^2}\right), \quad \gamma = \dfrac{2E_0}{Pa^2b^2}, \qquad \delta = \dfrac{E_0\kappa}{Pa^2b^2}$

To solve for the stress function, as in Ref.[13] differentiate both sides of the governing equations w.r.t. U after considering the average value of the parameters involving e' in terms of e_0, the central plastic strain. The value of e_0 is determined for $U = 1$ [vide Ref.[27].

Compatible with boundary conditions, w may be assumed as a power series. A single term approximation with a Galerkin procedure yields quite satisfactory and excellent comparable results minimizing the computational difficulties. Let us assume w for the said structure with clamped immovable edges[13] as

$$w(U,t)=h\,W(U)\,\Psi(t)=h\,\Psi(t)\sum_{j=2}^{n}U^{j}\approx hU^{2}\Psi(t) \qquad (8.15)$$

With this value of w, from Eqn. (8.14), the first and the second integrals may be obtained and finally we obtain

$$\left\{(1-U\,)\frac{dF}{dU}\right\}=(1-\Omega\,)\left[\frac{1}{4}lU^{\,3}+\frac{1}{3}mU^{\,2}\right]$$
$$+\frac{d\Omega}{dU}\left[p\!\left(-\frac{9}{5}U^{\,5}+U^{\,4}+2U^{\,3}+16\,U^{\,2}\right)+\frac{1}{4}QU^{\,3}\right]$$
$$+C_{1}U+C_{2}$$

$$(8.16)$$

where

$$l=\frac{4\gamma h^{2}\Psi^{2}}{3},\quad m=h\,\delta\,\Psi\,,\quad p=(\gamma h^{2}\,\Psi^{2})/36\ ,\quad Q=\frac{h\,\delta\,\Psi}{3}$$

$$(8.17)$$

Considering the case for a clamped immovable edge condition we set the following conditions[13] on F

$$\left.\frac{dF}{du}\right|_{u=0}=0\quad and\quad \left.\left\{(1-u)\frac{d^{2}F}{du^{2}}-2(1-v)\frac{dF}{du}\right\}\right|_{u=0}=0$$

which makes both C_1 and C_2 to be zero and Eqn. (8.16) thus reduces to

$$\left\{(1-U)\frac{dF}{dU}\right\}=(1-\Omega)\left[\frac{1}{4}lU^{3}+\frac{1}{3}mU^{2}\right]+\frac{d\Omega}{dU}\left[p\!\left(-\frac{9}{5}U^{5}+U^{4}+2U^{3}+16U^{2}\right)+\frac{1}{4}QU^{4}\right]$$

$$(8.18)$$

Our next step is to solve Eqn. (8.13). To avoid the integral in the term involving w_{tt} let us differentiate both sides of this equation w.r.t. U, to obtain

$$(1-\Omega)_0\left[(1-U)^2\frac{d^4w}{dU^4}-4(1-U)\frac{d^3w}{dU^3}+2\frac{d^2w}{dU^2}\right]$$

$$-[\frac{d\Omega}{dU}]_0\frac{d}{dU}\left[(1-U^2)\frac{d^2w}{dU^2}-2\frac{P^*}{P}(1-U)\frac{dw}{dU}\right]-$$

$$+\alpha\frac{d}{dU}\left\{(1-U)\left(\frac{dw}{dU}\frac{dF}{dU}\right)\right\}+\frac{d}{dU}\left\{\beta(1-U)\frac{dF}{dU}\right\}$$

$$-\frac{q}{2D_0P}+\frac{\rho h}{2D_0P}w_{tt}=0$$

$$(8.19)$$

On substitution of assumed values of w and $\left\{(1-U)\dfrac{dF}{dU}\right\}$ from Eqns (8.15) and (8.18) in Eqn. (8.19), the error function is obtained, which on application of Galerkin process yields the following time differential equation as

$$\rho h^2\ddot{\Psi}+\alpha_1\Psi+\alpha_2\Psi^2+\alpha_3\Psi^3=Q^*\qquad(8.20)$$

where,

$$\alpha_1=[(10D_0P)\left(\frac{4}{3}+\frac{1}{5}\beta\delta\right)+\left(\frac{1}{3}+\frac{2}{3}\frac{P^*}{P}+\frac{1}{18}\beta\delta\right)\left(\frac{d\Omega}{dU}\right)_0]h$$

$$\alpha_2=(10D_0P)\left[(1-\Omega)_0\left(\frac{4}{9}\alpha\delta+\frac{2}{9}\beta\lambda\right)+\left(\frac{5}{43}\alpha\delta+\frac{94}{945}\beta\gamma\right)\left(\frac{d\Omega}{dU}\right)_0\right]h^2$$

$$\alpha_3=\frac{10}{21}D_0P(\alpha\gamma)\left[10(1-\Omega)_0+\frac{1889}{90}\left(\frac{d\Omega}{dU}\right)_0\right]h^3$$

$$(8.21)$$

$$Q^*=(5/3)q\qquad(8.22)$$

Equation (8.20) with Eqns. (8.21, 8.22) can be used for Static and Dynamic analysis of Shells.

8.3   Verification of the Resulting Governing Equations

8.3(i)   Nonlinear Analysis of Elastic Plates

Set $\Omega = 0$, when (8.19) and (8.20) turn out exactly equal to be those as previously obtained in Ref. [28]. So, we may omit any further discussion on it.

8.3(ii)   Linear Free Vibration

Set $\alpha_2, \alpha_3 and$ Q^* equals to zero, when the linear frequency ω_{LP}^* is given by

$$\omega_{LP}^{*\,2} = \frac{Eh^2 P}{8\rho}\left[\frac{(320/3)}{12(1-v^2)} + 4\left(\frac{2\gamma_1}{h}\right)^2\right] + \frac{E_0 h^2 P}{72\rho}\left\{\frac{2}{(1-v^2)} + \frac{4}{(1-v^2)}\frac{P^*}{P}+\left(\frac{2\gamma_1}{h}\right)^2\right\}\left(\frac{d\Omega}{dU}\right)_0$$

$$(8.23)$$

here $\gamma_1 = \left(\frac{\kappa}{Pa^2 b^2}\right)$ and $\frac{2\gamma_1}{h} = \gamma^*$ represents the measure of shallowness of the shell. For Elastic shells, the corresponding linear frequency ω_{LE}^* is obtained from (8.23) by setting $\Omega = 0$, as

$$\omega_{LE}^2 = \frac{Eh^2 P}{8\rho}\left[\frac{(320/3)}{12(1-v^2)} + 4\left(\frac{2\gamma_1}{h}\right)^2\right]$$

which exactly equals to those obtained in Refs. [10,24] after making some simplification or with a little rearrangement of terms involving the parameters.

8.3(iii)   Nonlinear Free Vibration

Set $Q^* = 0$ in Eqn.(8.20) to get

$$\ddot{\Psi} + \frac{\alpha_1}{\rho h^2}\Psi + \frac{\alpha_2}{\rho h^2}\Psi^2 + \frac{\alpha}{\rho h^2}\Psi^3 = 0$$

$$(8.24)$$

Eqn.(8.24) is a familiar form of time differential equation and for which the frequency ratio (Nonlinear to Linear) is given by

$$
\begin{aligned}
&L_1\left[\begin{array}{l}
(1-\Omega)_0\left\{\dfrac{2}{3(1-\nu^2)}+\dfrac{3}{10}\gamma*^2\right\} \\[2mm]
+\dfrac{1}{2(1-\nu^2)}\left\{\dfrac{1}{3}+\dfrac{2}{3}\dfrac{L_3}{L_2}+\dfrac{1}{6}(1-\nu^2)\gamma*^2\right\}\left(\dfrac{d\Omega}{dU}\right)_0
\end{array}\right]\Psi \\[4mm]
&+L_1\gamma*\left[(1-\Omega)_0\left\{\dfrac{16}{3L_2}+\dfrac{4}{3}\right\}+\left\{\dfrac{60}{43L_2}+\dfrac{188}{315}\right\}\left(\dfrac{d\Omega}{dU}\right)_0\right]\Psi^2 \\[4mm]
&+\dfrac{L_1}{L_2}\left[(1-\Omega)_0\dfrac{160}{7}+\dfrac{1889}{1890}\left(\dfrac{d\Omega}{dU}\right)_0\right]\Psi^3=\dfrac{qa^4}{E_0h^4}
\end{aligned}
$$

,

$$
\frac{\omega^*_{PNL}}{\omega^*_{PL}}=\left[1+\left\{\frac{3}{4}\frac{\alpha_3}{\alpha_1}-\frac{5}{6}\left(\frac{\alpha_2}{\alpha_1}\right)^2\right\}\left(\frac{A}{h}\right)^2\right]^{1/2}
$$

.

(8.25)

where *(A/h)* represents the relative amplitude.

8.3(iv) Static Analysis

Set $\ddot{\Psi}=0$ in Eqn.(8.20) , when Ψ represents the maximum deflection. On substitution of the values of the coefficients in the transformed equation the following equation depicts the load-deflection relationship with a little bit of simplifications as

(8.26)

$$
\begin{aligned}
&L_1\left[\begin{array}{l}
(1-\Omega)_0\left\{\dfrac{2}{3(1-\nu^2)}+\dfrac{3}{10}\gamma*^2\right\} \\[2mm]
+\dfrac{1}{2(1-\nu^2)}\left\{\dfrac{1}{3}+\dfrac{2}{3}\dfrac{L_3}{L_2}+\dfrac{1}{6}(1-\nu^2)\gamma*^2\right\}\left(\dfrac{d\Omega}{dU}\right)_0
\end{array}\right]\Psi \\[4mm]
&+L_1\gamma*\left[(1-\Omega)_0\left\{\dfrac{16}{3L_2}+\dfrac{4}{3}\right\}+\left\{\dfrac{60}{43L_2}+\dfrac{188}{315}\right\}\left(\dfrac{d\Omega}{dU}\right)_0\right]\Psi^2 \\[4mm]
&+\dfrac{L_1}{L_2}\left[(1-\Omega)_0\dfrac{160}{7}+\dfrac{1889}{1890}\left(\dfrac{d\Omega}{dU}\right)_0\right]\Psi^3=\dfrac{qa^4}{E_0h^4}
\end{aligned}
$$

where

$$L_1 = \left[3\left(\frac{a}{b}\right)^4 + 2\left(\frac{a}{b}\right)^2 + 3 \right], \quad L_2 = \left[3\left(\frac{a}{b}\right)^2 + 2 + \left(\frac{b}{a}\right)^2 \right], \quad L_3 = \left[\left(\frac{a}{b}\right)^2 + 2v + \left(\frac{b}{a}\right)^2 \right]$$

The results shown in Sec. 8.3 (iii) and (iv) for nonlinear vibration and static analysis cannot be verified to the authors knowledge there is no such available results in the literature.

8.4   Concluding Remarks

In conclusion it may be said that:

1. The application of constant deflection contour method is justified and appears to be simpler than existing methods.
2. The present method offers a new approach to deal with problems involving large amplitude vibrations of plates and shallow shells.
3. The significance of this method lies in the fact that the basic governing equations reduce to a set of ordinary differential equations.
4. The present method can be applied to study static as well as dynamic behaviour of plates and shells having arbitrary shaped boundaries for which other methods may fail. One point must be mentioned here that the present method relies heavily with the assumed form of the contour $U(x,y) = constant$ close to the correct equation. However, in most practical cases of interest they are defined.
5. The application of polynomial expressions for the deflection and the stress functions in conjunction with the Galerkin procedure appears to produce excellent results.
6. The comparison of results shows that even a moderately approximated assumption for the expression of the deflected function yields results close to the literature results in all cases, sometimes better than those obtained from other methods. These results appear to be more accurate for the linear cases. In the case of nonlinear vibration, the results compare very well with those previously obtained. Additionally, the load deflection relations coincide very closely with that of Way [28].
7. This also offers potential for analyzing the nonlinear response of complex structures. The use of Conformal mapping technique restricts

one in finding the proper mapping function. Such problems are currently under investigation and the results will be published in due course. However, a simple approach to such structures has been presented in APPENDIX.

In conclusion it may be accepted that the CDC method may become a simple tool to deal with the problem of nonlinear vibration of plates and shells of arbitrary shapes.

REFERENCES

1. A. W. Leissa, *Vibration of Plates*, NASA SP-160, 1969.
2. Th. V. Kármán, *Festigktsprobleme im Maschinenbau, Encyklopädie der Mathematischen Wissenchaften*, Vol.IV, No. 4, pp.311-385, 1910.
3. G. Herrmann, "Influence on large amplitudes on flexural motions of elastic plates," *NACA Tech. Note* 3578, 1955.
4. B. R. El-Zaouk and C. L. Dym *Jl. of Sound and Vivration* **31**,89-103, Nonlinear Vibrations of Orthotropic Doubly-Curved Shalow shells, 1973
5. J. Nowinski, Nonlinear transverse vibrations orthotropic cylindrical shells, *AIAA*, 1(3), 617-620, 1963.
6. W. Leissa and A. S. Kadi, Curvature effects on shallow shell Vibrations, *Jl. of Sound and Vivration*, Vol.16, 1971, 173-187.
7. D. L. Hill, and Mazumdar, J, A study of the thermally induced large vibrations of viscoelastic plates and shallow shells, Jl. Sound and Vibration, 116(2), 323-337, 1987.
8. J. Mazumdar,A method for solving problems of elastic plates of arbitrary shapes, J. Aust. Math. Soc., Vol. 11,1970, 95-112.
9. J. Mazumdar , Transverse vibration of elastic plates by the method of constant deflection lines, *J. Sound Vib.*, **18**, 1971, 147-155.
10. R. Jones and J. Mazumdar, Transverse vibrations of shallow shells by the method of constant-deflection contours, *J. Acoust. Soc. Am., Vol.56, No.5, November* 1974. 1487-1492.
11. J. Mazumdar, Transverse vibration of membranes of arbitrary shape by the method of constant deflection contours, J. Sound Vib., 27, 1973, 47-57.

12. M. M. Banerjee, A new approach to the nonlinear vibration analysis of plates and shells, trans.14th Intl. Conf. On Struc. Mech. In Reactor Tech., (SMIRT-13), Divn.B, Paper No. 247 (Lyon, France–1997).
13. M.M.Banerjee and G.A.Rogerson., On the application of the constant deflection-contour method to nonlinear vibrations of elastic plates. *Archive of Applied Mechanics*. (72), pp 279-292. 2002.
14. R. Jones, J. Mazumdar and Chiang, F. P., Further studies in the application of the method of constant deflection lines to plate bending problems, *Intl. J. Eng. Sc.*, 13, 423-443, 1975.
15. J. Mazumdar, Buckling of elastic plates by the method of constant deflection lines, *J. Aust.Math. Soc.*, 13, 91-103, 1971.
16. D. Bucco and J. Mazumdar, Vibration analysis of plates of arbitrary shape-A new approach *J. Sound Vib.*, 67(2), 253-262, 1979.
17. F. Elsbrand, A. W. Leisssa, Free vibration of a rectangular plate clamped on three edges and free on the fourth edge, Developments on Theoretical and Applied Mechanics, Vol. 8,(Ed. I)Frederick, Pergamon Press –Oxford & New York, 1970.
18. L. V. Kantorovich and V. I. Krylov, *Approximate Methods of Higher Analysis*, P. Noordhoff, Ltd., Gronigen, 1964 (translated from Russian text).
19. C. P. Vendhan and Y. C. Das, "Application of Rayleigh-Ritz and Galerkin Methods to Nonlinear Vibration of Plates," Jl. Sound and Viv.,39(2), pp. 147-157, 1975.
20. C. P. Vendhan, " An investigation into nonlinear vibrations of thin plates," *Int. Jl. Nonliear Mech.*, Vol. 12, pp. 209-221, 1977.
21. S. H. Crandal, *Engineering Analysis*, New York, McGraw-Hill Book Co.,Inc. 1956.
22. C. W. Bert, "Nonlinear vibration of a Rectangular plate arbitrarily laminated of anisotropic material,"Trans. ASME, vol. 95, (June 1973) pp.452-457, 1973.
23. R. G. Anderson, Irons, B. M. and Zienkiwicz, O. C., Vibration and stability of plates using finite elements, *Intl. J. Solids and Struc.*, 4, 1978, 1031-1055.
24. E. Reissner, On axi-symmetrical vibrations of shallow spherical shells,Quar. Appl. Math., 13(3), 1955, 279-290.
25. S. Timoshenko and S. Woinowski-Krizer, *Theory of Plates and Shells*, 2nd Edition, New York 1959, McGraw-Hill Book Co., p.410, Table 82.

26. A.A. Ilyushin, Plasicity (in Russian) OGIZ. G.I.T.T.L., Moscow-Leningard; (in French), Paris1956, Ed. Eyrolles

27. J. Mazumdar and R. K. Jain, Elastic-Plastic Analysis of Plates of Arbitrary Shape – A New Approach, InternationalJl. of Plasticity, vol.5, 463-475, 1989.

28. S. Way, "Bending of Circular Plates with Large Deflection" Trans ASME, 56, 1934, 627-636.

SOME ADDITIONAL REFERENCES

A1. J. Mayers and B. G. Wrenn *Developments in Mechanics, Proceedibgs of the Tenth Midwestern Mechanics Conference* **4**, 819. On the nonlinear free vibration of thin cylindrical shells. New York: Johnson Publishing Co. 1967.

A2. D. A. Evensen and R. E. Fulton 1965 *International Conference onDynamic Stability of Structures Evanston, Illinois (Edited by G. Herrmann),* Some studies on the nonlinear dynamic response of shell-type structures,1965.

A3. M. M. Banerjee, P. Biswas, and, S. Sikder, Temperature effect on the dynamic response of spherical shells, SMIRT-12 Trans., Vol. B, Paper No. B06/3, 1993, 159- 163.)

ACKNOWLEDGEMENT

The authors are most grateful to Dr. A. Mernone of The University of Adelaide for the final form of editing and typesetting of the manuscript.

APPENDIX

Application of the Conformal Mapping Technique

Very little have been investigated for the vibration analysis with the help of conformal mapping technique, which, of course, will be much help to the designers in dealing with some special structures. The authors have tried to make an humble attempt to set forth a preliminary approach to this technique.

A survey of literature on the nonlinear vibration analysis reveals that though several investigations have been made on thin elastic plates having boundaries in the form of curves expressible by the common coordinate systems (Cartesian or Polar) yet apparently there is dearth of literature on the pertinent analysis of large amplitude vibration of plates having irregular boundaries not natural to such coordinate systems. However, Laura and others[C4-C8] presented a very convenient tool using the conformal technique to transform the given domain into a simpler one, i.e., onto a unit circle for such complicated boundaries. Yet those problems are confined to linear in character only. The first author of the present paper has made some honest attempts to extend this technique to plates exhibiting large deflections[C12] and vibrating at large amplitudes[C15]. Introducing the complex coordinates $z = x + iy$, $\bar{z} = x - iy$ in conjunction with the concept of the present CDC-Method the basic governing equations will take the following forms

$$\frac{\partial^4 w}{\partial z^2 \partial \bar{z}^2} = -\frac{h}{4D}\left[\phi_{zz} w_{\bar{z}\bar{z}} - 2\phi_{z\bar{z}}w_{z\bar{z}} + \phi_{\bar{z}\bar{z}}w_{zz}\right] + \frac{q - \rho h w_{tt}}{16D} \quad (c1)$$

$$\frac{\partial^4 \phi}{\partial z^2 \partial \bar{z}^2} = \frac{E}{4}\left[w_{zz}\, w_{\bar{z}\bar{z}} - w_{z\bar{z}}^2\right] \quad (c2)$$

On transformation to u-variables as performed earlier the above equations will further reduce to

$$16D\sum_{i=1}^{4} l_i \frac{d^{5-i}w}{du^{5-i}} = -4h\left[l_5 \frac{d}{du}(\frac{dw}{du}\frac{d\phi}{du}) + 2l_6 \frac{dw}{du}\frac{d\phi}{du}\right] + q - \rho h w_{tt}$$
$$(c3)$$

$$16 \sum_{i=1}^{4} l_i \frac{d^{5-i}\phi}{du^{5-i}} hF \ (t) = 4 E \left[l_5 \frac{d^2 w}{du^2} \frac{dw}{du} + l_6 (\frac{dw}{du})^2 \right] \quad \text{(c4)}$$

where

$$l_1 = u_z^2 u_{\bar{z}}^2, \qquad l_2 = (u_z^2 u_{\bar{z}\bar{z}} + 4u_z u_{\bar{z}} u_{z\bar{z}} + u_{zz} u_{\bar{z}}^2),$$
$$l_3 = (2u_{z\bar{z}}^2 + 2u_z u_{z\bar{z}\bar{z}} + 2u_{\bar{z}} u_{zz\bar{z}} + u_{zz} u_{\bar{z}\bar{z}}), \quad l_4 = u_{zz\bar{z}\bar{z}},$$
$$l_5 = (u_z^2 u_{zz} - 2u_z u_{\bar{z}} u_{z\bar{z}} + u_{\bar{z}\bar{z}} u_z^2),$$
$$l_6 = (u_{zz} u_{\bar{z}\bar{z}} - u_{z\bar{z}}^2).$$

$$\text{(c5)}$$

Assuming that there is a functional relation $z = f(\varsigma)$ which maps the given shape of the plate in the z-plane into a unit circle in the ζ-plane, one obtains similar equations (c3-c4). The only difference is that the l_i's are substantially changed to, say l_i' dependent on $\varsigma, \bar{\varsigma}$, as,

$$16 \ D \sum_{i=1}^{4} l_i' \frac{d^{5-i} w}{du^{5-i}} = -4h \left[l_5' \frac{d}{du} (\frac{dw}{du} \frac{d\phi}{du}) + 2 l_6' \frac{dw}{du} \frac{d\phi}{du} \right]$$
$$+ q - \rho h w_{tt}$$

$$\text{(c6)}$$

$$16 \sum_{i=1}^{4} l_i' \frac{d^{5-i}\phi}{du^{5-i}} hF \ (t) = 4 E \left[l_5' \frac{d^2 w}{du^2} \frac{dw}{du} + l_6' (\frac{dw}{du})^2 \right]$$
$$\text{(c7)}$$

where l_i' are given by

$$l_1' = (u_\varsigma^2 u_{\bar{\varsigma}}^2 z' \bar{z}' / z'^3 \bar{z}'^3), \quad m' = z'^3 \bar{z}'^3,$$
$$l_2' = \{ u_\varsigma^2 (u_{\varsigma\bar{\varsigma}} \bar{z}' - u_{\bar{\varsigma}} \bar{z}'') z' + 4u_\varsigma u_{\bar{\varsigma}} u_{\varsigma\bar{\varsigma}} z' \bar{z}'$$
$$+ u_{\bar{\varsigma}}^2 (u_{\varsigma\varsigma} z' - u_\varsigma z'') \bar{z}' \} / m'$$
$$l_3' = \{2 u_{\varsigma\bar{\varsigma}}^2 z' \bar{z}' + 2 u_\varsigma (u_{\varsigma\bar{\varsigma}\bar{\varsigma}} \bar{z}' - u_{\varsigma\bar{\varsigma}} \bar{z}'') z' + 2 u_\varsigma (u_{\varsigma\varsigma\bar{\varsigma}} z' - u_{\varsigma\varsigma} z'') \bar{z}' \} / m'$$

$$+ \{(u_{\varsigma\varsigma} z' - u_\varsigma z'')(u_{\bar{\varsigma}\bar{\varsigma}} \bar{z}' - u_{\bar{\varsigma}} \bar{z}'') + l_4 \} / m'$$
$$l_4' = u_{\varsigma\varsigma\bar{\varsigma}\bar{\varsigma}} z' \bar{z}' - u_{\varsigma\varsigma\bar{\varsigma}} z' \bar{z}'' - u_{\varsigma\bar{\varsigma}\bar{\varsigma}} z'' \bar{z}' + u_{\varsigma\bar{\varsigma}} z'' \bar{z}''$$

$$l'_5 = \{ u^2_{\bar{\varsigma}}(u_{\varsigma\bar{\varsigma}}\,\overline{z}' - u_{\bar{\varsigma}}\,\overline{z}'')z' - 2\,u_{\varsigma}u_{\bar{\varsigma}}u_{\varsigma\bar{\varsigma}}\,z'\,\overline{z}' +$$

$$u^2_{\bar{\varsigma}}(u_{\varsigma\varsigma}\,z' - u_{\varsigma}\,z'')\overline{z}'\ \}/\,m'$$

$$l'_6 = \left[(u_{\varsigma\varsigma}\,z' - u_{\varsigma}\,z'')(u_{\bar{\varsigma}\bar{\varsigma}}\,\overline{z}' - u_{\bar{\varsigma}}\,\overline{z}'') - u^2_{\varsigma\bar{\varsigma}}\,z'\overline{z}'\right]/\,m' \tag{c8}$$

and $z = f(\varsigma)$, $\overline{z} = f(\overline{\varsigma})$, $z\,\overline{z}$ and $\varsigma(=re^{i\theta})$, $\overline{\varsigma}$ are complex conjugate in their complex planes, respectively.

Method of Solution

It is not possible to obtain an exact solution for the transformed equations (c6, c7) and hence at first we have to search for a suitable expression for the mapping function $z = f(\varsigma)$ and the deflection function $W(\zeta,\overline{\zeta})$ compatible with the prescribed boundary conditions.

While employing a conformal technique one must be careful that a proper mapping function for the contour is known. Sometimes the mapping function may be expressed in a power series. For example, in the case of a regular polygon, the power series may be expressed in the form (See Ref.C14)

$$z = f(\zeta) = a_p A_s \sum_{n=0}^{\infty} a_n \zeta^{1+ns} \tag{c9}$$

where :

$a_p = apothem$,

$$A_s = {1}\Big/{\int_0^1 (1 - u^s)^{2/s}\,du}$$

$$a_0 = 1, \qquad a_n = \frac{(2 + s)(2 + 2s)\ldots\ldots[\ 2 + (n-1)s]}{s^n n!(1 + ns)}(-1)^n$$

$$\tag{c10}$$

s = number of axis of symmetry, A_s are known as the mapping function coefficients.

Laura has prescribed the values of $z = f(\varsigma)$ and A_s for some structures of arbitrary shapes which may be given here for ready reference. For the deflection function a general expression has also been reported in Ref.25 in the form

$$z = f(\varsigma) = a_p \, \lambda \left(\zeta + \mu\zeta^{m+1}\right) \tag{c11}$$

$$W(\zeta,\bar{\zeta}) = \sum_{n=1}^{N} b_n \left[1 - \left(\zeta\bar{\zeta}\right)^n\right] \tag{c12}$$

Table-6 Shapes and their respective values for parameters s, A_s, m, λ, μ

Shapes	s	A_s	m	λ	μ
Triangular plate	3	1.1352			
Square plate	4	1.0788			
Pentagon	5	1.0516			
Hexagon	6	1.0376			
Heptagon	7	1.0279			
Octagon	8	1.0220			
Circular				1.00	0.00
Circular plate with two flat sides			2	0.9224	0.00
Elliptic plate $\dfrac{x^2}{4/3} + \dfrac{y^2}{4/5} = 1$				0.99	0.101
Square plate with rounded corner			4	25/45	-.04

Some results obtained using conformal mapping technique

Considered here vibration of a square plate of side 2a for which
$$z = f(\varsigma) = a(1.08\,\zeta - 0.108\,\zeta^5 + 0.045\,\zeta^9)$$
with

$$W\left(\zeta,\bar{\zeta},t\right) = \sum_{k=1}^{k} b_k \psi_k(\zeta,\bar{\zeta})F(t) = \sum_{k=1}^{k} b_k[1 - (\zeta,\bar{\zeta})^2\,F(t)$$

and following the method described earlier, the time differential equation is in excellent agreement with that of Yamaki[C1].

Moreover, the conformal approach has been successfully made in the first author's note[C15] when the results of large deflection problem was found to be in close agreement with known available results.

From the results given above for static and dynamic analysis of plates vibrating at moderately large amplitude, it appears that the application of the present method is justified. For the linear analysis the results obtained by this method are closer to exact results than those obtained by other methods. From Table-2 it is clear that the nonlinear effect is more predominant in the case of an annular plate. Considering the static nonlinear case, the values of the load parameter are close to those given by Way[C5], and even better than those obtained by Yamaki[C1]. Also the results are in excellent agreement with those given in Ref. [C6].

The results for static deflection of annular plates cannot be compared for non-availability of results in the literature. However, for the linear case the result can be compared with those given in Ref. [C4],

$$W_{max}/h = 0.0575 \frac{qa^4}{Eh^4} \text{ (Ref.[C4])}, \quad W_{max}/h = 0.0566 \frac{qa^4}{Eh^4} \text{ (Present)}$$

Moreover, in the nonlinear static case the results for RCP are even better than those of Yamaki[C1], and that the present results are closer to those of Way[C5].

In conclusion it may be stressed that the application of constant deflection contour method appears to be simpler than existing methods. The most important point with regard to the present method is that it can be applied to study static problems, as well as dynamic behaviour of structures having complex boundaries for which other methods may fail. One point must be mentioned here about the present method that it completely relies on the judicial selection of the contour $U(x,y) = constant..$ However, in most practical cases of interest this poses no problem.

The application of polynomial expressions for the deflection and the stress functions in conjunction with the Galerkin procedure appears to produce excellent results. These results appear to be more accurate for the linear cases. In the case of nonlinear vibration, the results compare quite well with those previously obtained [C1, C6].

Additionally, the load deflection relations coincide very closely with that of Way[C5]. The present paper offers a new approach to deal with problems involving large amplitude vibration of plates. This also offers potential for analyzing the nonlinear response of complex structures. The use of Conformal mapping technique restricts one in finding the proper mapping function. However, Laura *et al*[C10] have provided means to evaluate them and listed some of the mapping functions for regular polygonal shapes.

REFERENCES (for Appendix)

C1.   Yamaki, N., Influence of large amplitude on flexural vibrations of elastic plates, ZAMM, 41, pp.501-510, 1961.

C2.   Mazumdar, J., A method for solving problems of elastic plates of arbitrary shapes, *J. Aust. Math. Soc.*, 11, 1970, 95-112.

C3.   Banerjee, M. M., A new approach to the nonlinear vibration analysis of plates and shells,Trans.14th Intl. Conf. On Struc. Mech. In Reactor Tech., (SMIRT-13), Divn. B, Paper No. 247 (Lyon, France–1997).

C4.   Timoshenko, S.P. and Woinowisky-Krieger., *Theory of plates and shells*, 2nd Edn., McGraw-Hill, New York, 1959.

C5.   Way, S., Bending of circular plates with large deflection, *Trans. ASME*, 56, 1934, 627-636.

C6.   M.M.Banerjee and G.A.Rogerson., An application of the constant contour deflection method to non-linear vibration. *Archive of Applied Mechanics*. (72), pp. 279-292. 2002.

C7.   Laura, P.A. A. and Shahady, P., Complex variable theory and elastic stability problems, *Jl. Engg. Mech. Divn.,ASCE*,95,No.EML. Proc., paper 6386, pp.59-67, 1969.

C8.   Pombo, J. L.,Laura, P.A.A., Gutierrez,R.H. and Steinberg, D.S., Analytical and experimental investigation of the free vibration of clamped plates of regular polygonal shapes carrying concentrated masses.,*Jl. Sound & Vib.*,55(4). pp521- 532, 1977.

C9.   Laura, P.A.A. and Rumanelli, E.,Determination of eigen values in a class of wave-guides of doubly connected cross section. Jl. Sound & Vib., 26(3), pp.395-400, 973.

C10. Laura, P.A.A. and Chi, M.,An application of conformal mapping to a three dimensional unsteady heat conduction problem., Aeronautical Quarterly, 16, pp.221-230, 1965.

C11. Leissa, A.W., Laura, P.A.A. and Gutierrez,R.H, Transverse vibrations of circular plates having non-uniform edge constraints, Jl. Acous. Soc. Am., 66(1), July 1979, pp.180-84, 1967.

C12. Banerjee, M. M., Note on the large deflections of irregular shaped plates by the method of conformal mapping, Trans. ASME, June 1976, pp.356-57, 1976.

C13. Chakrabarty,S and Banerjee,M.M., Large amplitude vibration of plates of arbitrary shape With uniform thickness,Proc.4th ICOVP, Jadavpur University

(INDIA), Nov.27-30, pp.85-89, 1999.

C14. Laura, P. A.A., and Luisoni, L.E., Proc. of the Soc. Of Experimental Stress Analysis Vol.33pp.279-280. Discussion on the paper "An analytical and experimental study of vibrating equilateral triangular plates", 1976.

C15. Banerjee, M. M. and Das,J. N., Use of conformal mapping technique on dynamic response of plate structures, Proc. of 1st ICOVP, A.C.College, Jalpaiguri,) October 20-24, l990, pp.129 – 31, 1990.

--- o---